AF134489

Microbiology Research Advances

nova
science publishers

www.novapublishers.com

Microbiology Research Advances

Endophytic Fungi: Biodiversity, Antimicrobial Activity and Ecological Implications
Qiang-Sheng Wu, PhD (Editor)
Ying-Ning Zou (Editor)
Yong-Jie Xu, PhD (Editor)
2021. ISBN: 978-1-68507-354-1 (Softcover)
2021. ISBN: 978-1-68507-442-5 (eBook)

Pathogenic Bacteria: Pathogenesis, Virulence Factors and Antibacterial Treatment Strategies
Keith D. Watts (Editor)
2021. ISBN: 978-1-68507-422-7 (Softcover)
2021. ISBN: 978-1-68507-424-1 (eBook)

Biofilms: Advances in Research and Applications
Shane Rowland (Editor)
2021. ISBN: 978-1-68507-062-5 (Hardcover)
2021. ISBN: 978-1-68507-089-2 (eBook)

Antimicrobial Peptides: Functions, Mechanisms of Action and Role in Health and Disease
Paresh Chandra Ray, PhD (Editor)
2021. ISBN: 978-1-68507-005-2 (Hardcover)
2021. ISBN: 978-1-68507-082-3 (eBook)

Biopharmacological Activities of Medicinal Plants and Bioactive Compounds
Dr. Ajeet Singh, PhD (Editor)
Dr. Navneet (Editor)
2021. ISBN: 978-1-53619-959-8 (Hardcover)
2021. ISBN: 978-1-68507-056-4 (eBook)

More information about this series can be found at
https://novapublishers.com/product-category/series/microbiology-research-advances/

Arif Pandit and R. S. Sethi
Editors

Endotoxins and Their Importance

nova
science publishers
www.novapublishers.com

DOI: https://doi.org/10.52305/XDYQ9654

NOTICE TO THE READER

Library of Congress Cataloging-in-Publication Data

Names: Pandit, Arif, editor. | Sethi, R. S. (Ram Saran), editor.
Title: Endotoxins and their importance / Arif Pandit, PhD (editor), Assistant Director Research, Directorate of Research, SKUAST-Kashmir. J&K, India, R.S. Sethi, PhD (editor), Professor and Head, Department of Animal Biotechnology, College of Animal Biotechnology, Guru Angad Dev Vety and Animal Sciences University, Ludhiana, Punjab, India.
Description: New York : Nova Science Publishers, [2022] | Series: Microbiology research advances | Includes bibliographical references and index. |
Identifiers: LCCN 2022020330 (print) | LCCN 2022020331 (ebook) | ISBN 9781685078393 (paperback) | ISBN 9781685079130 (pdf)
Subjects: LCSH: Endotoxins.
Classification: LCC QP632.E4 E536 2022 (print) | LCC QP632.E4 (ebook) | DDC 615.9/5293--dc23/eng/20220613
LC record available at https://lccn.loc.gov/2022020330
LC ebook record available at https://lccn.loc.gov/2022020331

Published by Nova Science Publishers, Inc. † New York

Contents

Preface

Endotoxins, also known as LPS, constitute a significant component of Gram-negative bacteria's outer membrane that activates the innate immune system. Lipid A, a short core oligosaccharide, and the O-antigen polysaccharide make up LPS. Lipid A is the primary immune response trigger and is required for bacterial growth. To find a new antibacterial target, researchers must first understand endotoxin biosynthesis, virulence, and antibiotic resistance. Bacteria use biosynthetic pathways to resist antibiotics and survive many treatment regimes. The type of condition that develops and the risk associated with it can be influenced by the timing of an agent's exposure. Endotoxins are more common in the biomedical, livestock, agricultural, and food industries. The threat of endotoxin exposure in the environment has been discussed in this book. The pathogenic effects of endotoxin influenced by the host's age, family history, and immune function are the broader areas covered in this book. Further, the principles, practices, and current knowledge of endotoxins are examined in this book, which focuses on their structure, physical and chemical characterization, biosynthesis, virulence regulation, and presence in the environment. Students, clinicians, young scientists, professionals, and researchers can be sure that the facts you learn about endotoxins and how they can be used to treat human and animal diseases are updated with the latest research on the topic.

Arif Ahmad Pandit

RS Sethi

Chapter 1

Endotoxins: A Basic Introduction

M. P. S. Tomar[1,*], Chetna Mahajan[2], Geetika Verma[3], Rahul Nanotkar[4] and Deepak Sumbria[5]

[1]N.T.R. College of Veterinary Science, Gannavaram, India
[2]Department of Veterinary Biochemistry, College of Veterinary Science, Rampura Phul, Bathinda; Guru Angad Dev Veterinary and Animal Sciences University, Punjab, India
[3]Senior Demonstrator, Post Graduate Institute of Medical Education and Research (PGIMER) Chandigarh, India
[4]Department of Veterinary Pharmacology and Toxicology, N.T.R. College of Veterinary Science, Gannavaram, India
[5]Department of Veterinary Parasitology, College of Veterinary Science, Rampura Phul, Bathinda; Guru Angad Dev Veterinary and Animal Sciences University, Punjab, India

Abstract

Endotoxin is the harmful substance present inside the microorganism in the outer membrane. Chiefly gram-negative bacteria have Endotoxin. They are constituted by covalent bond interconnected lipopolysaccharides (LPS) having lipid (Lipid A), polysaccharide and O-antigen components. Regarding its positioning, it was noticed that Lipid A part is the inner component and is present in the cell membrane. It is followed by polysaccharide and then at last by O-antigen part.

LPS is an excellent immunological stimulant and has thermostable and hydrophobic properties. It is also present in a different environment, and its value can reach up to µg/m3. These toxins are secreted by vesicles and by the destruction of microorganisms. They act as antigens via Toll-like receptor-4, myeloid differentiating factor-2, CD14 cell surface protein, and cause pro-inflammatory cytokine storm from immune cells via internalization. Production of these toxins also causes endotoxemia

* Corresponding Author's Email: anatomistpdtr@gmail.com.

In: Endotoxins and their Importance
Editors: Arif Pandit and R. S. Sethi
ISBN: 978-1-68507-839-3

and fever, organ failure and septic shock. The chapter elaborates the history and basic introduction of endotoxins.

Keywords: endotoxins, gram negative bacteria, lipopolysaccharides, toxemia

Introduction

According to Beutler and Rietschel, 2003, the Endotoxin was first defined in 1892 by Richard Pfeiffer, who stated that the Endotoxin is a heat-stable, toxic substance released upon disruption of microbial envelopes. The term "Endotoxin" was introduced in the 19th century to describe an integral part of gram-negative bacteria's outer membrane-associated with proteins and other cell membrane-related components (Aderem and Ulevitch 2000). It was the primary factor for the pathophysiological outcomes associated with gram-negative infections. During the 18th century, most of the study was done on pyrogens, i.e., the factors which can induce fever and sepsis in unhygienic conditions, and it yielded a substance that was later termed as endotoxins (Aderem and Ulevitch 2000). The asymptomatic and nonclinical immunogenicity for these compounds can be seen in an individual, and an excellent immune response was seen when these were administered at a concentration less than one ng/mL (Suda et al. 2001).

Technical advancement in analytical chemistry and instrumentation helped reveal the chemical structures and biological properties of Endotoxin during the late 20th century (Wenqiong and Xianting, 2015). Gram-negative bacteria and their endotoxins that result in various pathophysiological outcomes were observed in vertebrates and non-vertebrate hosts. An endotoxin consists of Lipopolysaccharides, a well-documented pyrogen and is a vital part of the exterior cell wall of Gram-negative bacteria, e.g., *Escherichia coli, Salmonella spp.* etc.

In the normal life cycle of bacteria, the LPS is only secreted during cell division and in minute amounts. But any bactericidal activity (by antibiotics) or destructive reaction (by the host immune system) may result in the release of these chemicals and are capable of deadly outcomes (Aderem and Ulevitch 2000, Hoshino et al. 1999).

Endotoxin, synonymously called Lipopolysaccharide, is a naturally occurring part found in the outer membrane of all gram-negative bacteria. Ramachandran (2014) concluded that the Lipopolysaccharide consists of the lipid A part of fatty acids and disaccharide phosphates core. They contribute about 75% of the outer membrane of gram-negative bacteria that can cause a lethal shock.

Endotoxins are a non-concrete category of biomolecules that are released bacterial death and result in toxic effects, viz., fever, septic shock, multi-organ failure, and even mortality (Gutsmann et al. 2007). These immune responses oversee Lipid A, which is the active moiety of the LPS biochemical structure. Lipid A is the membrane anchor and single region of LPS molecules, composed of a hydrophilic, negatively charged bisphosphorylated diglucosamine backbone and a hydrophobic domain of six (*Escherichia coli*) or seven (*Salmonella*) acyl chains in amide and ester linkages. (Wenqiong and Xianting, 2015). Because of its unique structure, it is recognized by the innate immune system even at the picomolar concentrations, which is sufficient to trigger a macrophage to produce pro-inflammatory cytokines like TNF-α and IL1β and therefore oversees LPS molecules' biological function, specificity, and affinity to the relative proteins (Ramachandra, 2014; Wenqiong and Xianting, 2015).

Structure of Cell Wall of Gram-Negative Bacteria

Raetz and Whitfield (2002) concluded that in most Gram-negative bacteria, Endotoxin is one of the major components present in the cell wall's outer membrane. The envelope of Gram-negative bacterial cells is composed of the outer membrane (OM), the peptidoglycan (murein) and the inner membrane (IM). The outer membrane is unique to Gram-negative bacteria it serves as a protective structure. The lipid part of OM is highly asymmetric as its outer monolayer consists of a unique molecule known as lipid A in most of gram-negative bacteria. Aderem and Ulevitch (2000); Raetz and Whitfield (2002); Kothani et al. (1985); Galanos et al. (1985); Rietschel and Wespphal (1999) reported that the sn-l,2-diacylglycerol moiety of classical membrane phospholipids is absent in lipid A (Figure 1), and which is replaced by a 2 ,3-diacylglucosamine unit. The acyl chains linked to A's glucosamine backbone are 2-6 carbon atoms shorter and have an R-3-hydroxyl substituent from those present in glycerophospholipids. Lipid A's unique structure helps in specific outer membrane assembly and function roles, providing resistance to phospholipase. The periplasm, the gelatinous material present between the outer membrane and the IM has the enzymes necessary for a nutrient breakdown and binding proteins to help their transfer across the IM. Peptidoglycan is the periplasmic space that consists of alternating N-acetyl glucosamine (GlcNAc) and N-acetylmuramic acid (MurNAc) sugars which are cross-linked by short peptide bridges and keeps the osmotic pressure and cell structure (Holtje, 1998). The IM is a phospholipids bilayer like the cell membrane of eukaryotic cells and is permeable to lipophilic compounds. Numerous integral transmembrane, helical proteins and peripheral membrane proteins present in this layer handle transport, cell signalling and metabolic functions (Harald, 2001).

Kamio and Nikaido (1976), Galanos et al. (1985), Suda et al. (2001), Golenbock et al. (1991), Vogel et al. (1984) postulated that the Lipid A is the membrane anchor of Lipopolysaccharide (LPS) (Figure 1). Vogel et al. (1984) and Ancuta et al. (1996) mentioned that the 6' position of Lipid A in LPS is glycosylated with a nonrepeating oligosaccharide and is designated as the core. According to Galanos et al. (1985), Suda et al. (2001), Golenbock et al. (1991), Vogel et al. (1984) an eight-carbon sugar, namely 3-deoxy-D-manno-octulosonic acid (KDO), is directly linked to lipid A. In contrast, the other core sugars include L-glycero-D-manno-heptose, glucose, galactose, and N-acetylglucosamine (Galanos et al. 1985, Suda et al. 2001, Golenbock et al. 1991, Vogel et al. 1984, Dandstrom et al. 1992).

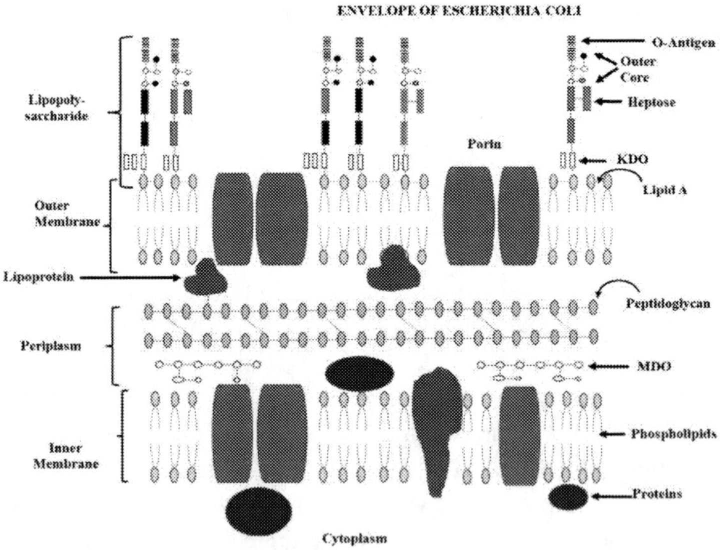

Figure 1.1. Schematic molecular representation of the E. coli envelope showing various components as lipopolysaccharide layer (O-antigen, Outer core, Heptose as bubbles and rectangles in lipopolysaccharide layer), outer membrane (composed of polar head groups of phospholipids, porin, lipoproteins and 3-deoxy-D-manno-octulosonic acid -KDO), periplasm (containing peptidoglycan), membrane-derived oligosaccharides (MDO), and inner membrane.

1.1. Biochemistry of Lipid A and Core Domains

Tobias et al. (1986), Takayama et al. (1983), Crowell et al. (1986) reported that the isolation, structure and conformation of Lipid A and its precursors have similar solubility properties as those of glycerophospholipids. However, an LPS molecule having a complete core and O-antigen does not have solubility as a monomer, both in water and in most organic solvents (Golenbock et al. 1991). However, the commercial LPS preparations form opalescent dispersions in water, and their structures and critical micellar concentrations have not been thoroughly characterized (Aderem and Ulevitch 2000). Garrett et al. (1997) found that aqueous phenol can be used to extract LPS from bacteria. While Galanos et al. (1985) and Ray et al. (1987) reported that mixtures of phenol, chloroform and petroleum ether were especially effective for extracting LPS lacking O-antigen. LPSs are arranged in a tightly packed structure in the outer membrane.

1.1.1. Structure of Lipopolysaccharide (LPS)

Three primary structural components of LPS are Lipid A, core oligosaccharide and O- specific antigens and the composition is like a lipid moiety known as Lipid A is coupled to an oligosaccharide known as the core oligosaccharide, which is further attached to a series of repeated subunits known as O-specific antigen (Tobias et al. 1986). Disaccharide is the backbone of Lipid A moiety with two negatively charged phosphate groups and as many as six acyl chains with 14–16 carbon atoms (Figure 1) (Ulevitch and Tobias 1999). Lipid A and O-region play critical components for endotoxin biological activity. Being amphiphilic, Lipid A shows hydrophobic and hydrophilic features, which may lead to aggregations in aqueous solutions resulting in three-dimensional supramolecular structures, known as micelles (Ulevitch and Tobias 1999, Shimazu et al. 1999). Bacterial serotype classification and evasion of serum complement attack are O-specific antigen-dependent. The O-specific antigen is used to classify the bacterial serotype and aids the bacteria in evading the attack of serum compliments from the host immune system (Parillo 1993; Van Deuren et al. 2001). Lipid A and polysaccharide components control the toxicity and immunogenicity, respectively. LPS components form the wall antigens (O antigens) of Gram-negative bacteria and can produce diversified inflammatory responses with complementary activation by alternative pathway (properdin) and may be a probable cause of Gram-negative bacterial infections pathology (Vishnupriya, 2015), while Lipid A is known to operate several *in vivo* and *in vitro* endotoxin effects. A molecular weight >100,000 D, LPS can regulate immune responses and cellular immune reactions. However, sufficient peptidoglycan (PGN) in commercial bacterial LPS Endotoxin activates antimicrobial peptides (Dziarski and Gupta, 2006). Endotoxins can stimulate immune responses at little activity while may lead to septic shock at elevated activities. Endotoxins released during the growth of Gram-negative bacteria simulates natural immunity. Endotoxins are believed to be heat stable but for some (boiling for 30 min does not destabilize Endotoxin), powerful oxidizing agents such as superoxide, peroxide, and hypochlorite (Vishnu Priya, 2018).

1.1.2. Toxicity of Endotoxin

LPS molecules of high chemical stability are released to the environment with the death of gram-negative bacteria cells. Endotoxemia, symptoms affecting the structure and function of many organs and cells, altered metabolic functions, elevated body temperature, hemodynamic changes, and septic shock can all result from an injured intestinal mucosa, which permits LPS molecules to enter human or animal blood. When LPS molecules access the human system by circulating through the liver, various inflammatory cytokines, such as tumour necrosis factor, interleukin-6, platelet-activating factor and so on, are overexpressed by the activation of the innate immune system and lead to systemic inflammatory response syndrome, which has been reported as the cause of death related to severe acute respiratory syndromes, cancers, large-area burns and acute peritonitis. The high mortality rate associated with endotoxin-induced shock stays a major clinical problem, especially in debilitated and immunosuppressed patients. Furthermore, Endotoxin is not eliminated when the sterilization process kills the organism; instead, the release of LPS takes place upon the death of cells (Wenqiong and Xianting, 2015).

Although the products may have been sterilized, the Endotoxin of the organism remains if gram-negative organisms were present before sterilization. Testing for this LPS in the finished products is an important part of ensuring the safety of the sterilized products, especially in biological products, medical devices, parenteral drugs, food, water security, etc.

1.1.3. Sepsis and Endotoxemia

Sepsis may be described as a medical condition with an overwhelming bacterial infection promoting an inflammatory state of the whole body, accompanied by fever, increased heart and respiratory rates (Ancuta et al. 1996). Severe sepsis with elevated levels of LPS is associated with a worse prognosis and more extended stays in the intensive care unit (Sandstrom et al. 1992). The diagnostic test for severe sepsis includes the measurement of endotoxins in blood and other bodily fluids (Sandstrom et al. 1992). Multiorgan failure is common, and the occurrence of acute liver damage is believed to be correlated to the presence of endotoxins (Girard et al. 2003, Darveau et al. 2004). Antibiotics damage the bacterial cell wall and release the endotoxins in the blood stream, thus declining the disease condition and are commonly used for treatment. (Heine et al. 2003, Ancuta et al. 1996). Momentary serum half-life and poor immunogenicity leads to the strenuous determination of LPS in human plasma and demands highly sensitive bioassays (Ancuta et al. 1996).

1.1.4. Periodontitis

Endotoxins are believed to play a role in developing chronic periodontitis, where the presence of LPS could result in delayed healing, inflammation, and reduced cell proliferation (Akira and Takeda 2004, Alipranitis et al. 1999). These endotoxins are released after the death or multiplication of bacteria. They have high virulence by stimulating the pro-inflammatory cytokine expression and, therefore, enhancing the host's immune response (Morra et al. 2015), which develops inflammatory reaction and bone resorption (Leonardo et al. 2004 and Nelson-Filho et al. 2011). Bacteria associated with periodontitis around teeth are also believed to contribute to failing implants (Andruciol et al., 2018). The process of implant loosening does not always reflect the progression of periodontal inflammation; nonetheless, the stimulation of bone resorption may be associated with gram-negative bacteria and LPS (Rietschel et al. 1994).

1.1.5. Aseptic Implant Loosening

The Association of endotoxins to the clinical setting of aseptic implant loosening is debatable. Still, increasing evidence indicates endotoxins as at least partially responsible for osteolysis and the loosening of the medical device (Rietschel et al. 1999, Nishijima et al. 19981, Nishijima et al. and Eaetz 1979, Trent 2004, Babinski et al. 2002, Raetz 1996). Aseptic implant loosening patients without any signs of microbial disease have shown LPS presence in the surrounding tissues (Whitfield 1995), and prophylactic antibiotics are used systemically, or it can reduce the incidence of aseptic loosening half. Also, wear particle inflammatory response, increase in the rate of foreign body reaction, discharge of pro-inflammatory cytokines and activation of macrophages are related to endotoxins (Raetz 1996, Schnaitman et al. 1993, Vaara 1993). Endotoxins can also play a key role in the detection of implant failure or impaired osteo-integration with the help of macrophages expressing endotoxins receptors, particularly TLR's, at one biomaterial interface detecting related cytokines with high sensitivity (Rietschel and Westphal 1999, Nishijima et al. 1981, Nikaido 1994, Coleman and Raetz 1988). It has also been reported that *in vitro* cell culture studies are significantly affected compared to *in vivo* settings (Raetz 1996).

There are many potential sources for endotoxins found in the periprosthetic tissue, including the bacterial biofilm on the surface of the implant, endotoxin contamination during the implant manufacture, and endotoxins derived from wear particles that can absorb LPS from systemic infections or the intestinal for a (Young et al. 1995). However, the removal of endotoxins was shown to have a more significant impact on in-vitro cell culture studies than in the in-vivo setting. This notion reinforces the call for strict monitoring of endotoxin contamination during the evaluation of wear debris in vitro models (Jackman et al. 1999).

1.1.6. Host-Microbe Interactions (Lipopolysaccharide Activity) in Invertebrates-Insects

LPS is present in the human tissue without any signs of bacterial infection, and the immune system can react to minute LPS levels as low as 1ng/ml presenting fever, hypotension and septic shock in elevated concentrations while may cause several chronic diseases in low levels (Aderem and Ulevitch 2000, Golenbock et al. 1991, Vogel et al. 1984). Also, elevated levels of endotoxins in the blood may further lead to severe consequences, *viz*. multiple organ failure, adult respiratory distress syndrome and disseminated intravascular coagulation (Aderem and Ulevitch 2000, Vogel et al. 1984).

Innate immunity is primarily responsible for insect's immunity (Hoffmann, 2003) and is known to have a highly effective defence involving both cellular and humoral responses against a variety of invaders *viz*. Gram-negative, Gram-positive, LPS, peptidoglycans and others (Hultmark, 1993). Also, the insect's defense system includes phagocytosis and encapsulation as biological responses (Lackie, 1988) of invaders with the help of hemocytes followed by the production of antimicrobial proteins (in the insect fat body), leading to its increased activity in the cell-free haemolymph of insect.

Strong immune activity was found in the interaction between *Galleria mellonella* and LPS. LPS acts as an immune stimulator for subsequent administration of Gram-negative bacteria. The high LPS-resistance may be explained by a very efficient detoxification mechanism which was reported to involve the binding of LPS by hemolymph lipophorins. This proved that LPS has pre-immune activation ability (Kato et al. 1994).

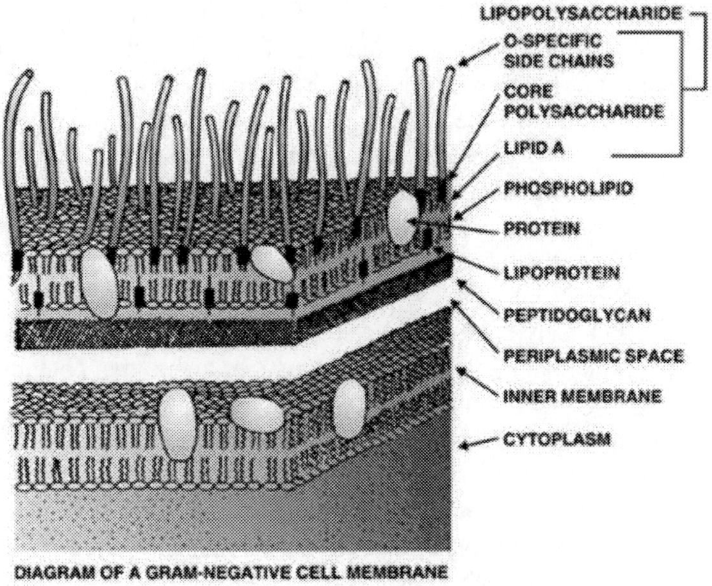

Figure 1.2. Illustration of a gram-negative cell membrane. https://www.horseshoecrab.org/med/endotoxin.html Illustration courtesy Charles River Endosafe, SC.

1.1.7. Biomaterial Endotoxin Testing

Endotoxins present a significant threat to human health in contamination in medical devices parenteral drugs (Kotani et al. 1985, Galanos et al. 1985, Bishop 2005). In this regard, certain safety limits of their contamination have been permitted depending upon application site in the medical device and is strictly regulated by Food and Drug Administration (FDA) (Kotani et al. 1985, Galanos et al. 1985, Bishop 2005) like The endotoxin levels below 0.5 EU/mL (20 EU/device) and 0.06 EU/mL (2.15 EU/device) for medical devices and devices in contact with the cerebrospinal fluid (Kotani et al. 1985, Galanos et al. 1985) are recommended. Presently, the kinetic limulus Limulus amebocyte lysate (LAL) method, which uses a lysate from a horseshoe crab to identify physiologically active LPS, and the rabbit pyrogen test, which measures the temperature rise in rabbits caused by endotoxins and other pyrogens (Bishop 2005, Robey et al. 2001), are used to measure the endotoxin content in the extract obtained by immersing the testing medical device in endotoxins free water for at least once in room temperature (Kotani et al. 1985, Bishop 2005). Also, endotoxins levels determined with the LAL method raises accuracy concerns (Kotani et al. 1985, Galanos et al. 1985, Rietschel et al. 1999, Moran et al. 1997, Robey et al. 2001, Pilione et al. 2004) with a limitation of determining cutoff values within a narrow concentration range and LPS content of the solution will depend on the properties of the biomaterial *viz.* surface properties, wettability, pH, and others (Galanos et al. 1985, Moranet al. 1997, Pilone et al. 2004). In addition to this, non-transparent and complex shapes further increase the problem several folds (Galanos et al. 1985). Such issues can overrule the actual biological response of the biomaterial (Galanos et al. 1985, Rietschel et al. 1999).

The amount of active and inactive forms of endotoxins can be assessed simultaneously, including gathering information about the contaminant's environmental origin (Preston et al. 2003). In contrast, the kinetic version of the assay introduces high error rates and may be more susceptible to influences from surface properties (Galanos et al. 1985). Another possibility to figure out the extent of endotoxin contamination is gas chromatography and mass spectrometry methods (Preston et al. 2003). This method, however, requires expensive equipment and specially trained personnel.

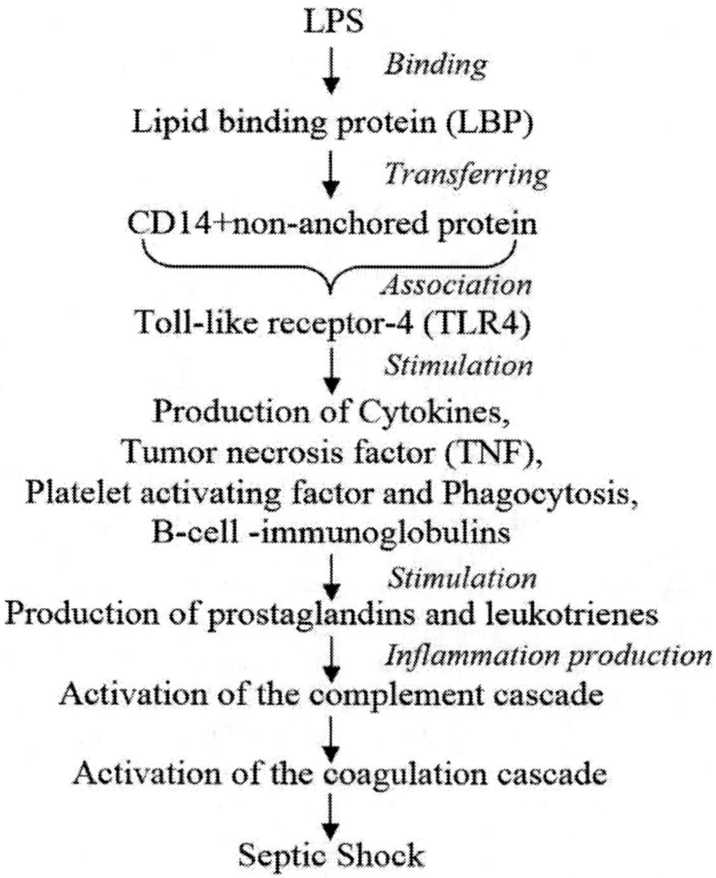

Figure 1.3. Schematic diagram of host pathogen interaction of endotoxins (Adapted from Vishnupriya, 2018)

1.1.8. Strategies for Endotoxin Removal in Biopharmaceutical Industries

There has been a tremendous growth in biotechnology industry for the generation of recombinant DNA for the production of various products viz. vaccines, antibodies, enzymes, proteins etc. For these purpose, Gram-negative bacteria are utilized widely and hence, there remain high chances of endotoxin contamination at any point of the process (Magalhaes et al. 2007). The removal of endotoxins is very important to obviate the chances of adverse side effects after administration of the product into human or animal. With the recent advances in this field, many methods have been devised for endotoxin removal viz. affinity chromatography, anion exchange chromatography (AEC), ultrafilteration, endotoxin-selective adsorbent matrices and two-phase extractions (Clarence et al. 2012). There has been increase in use of adsorbent matrices which are endotoxin selective for the removal of endotoxin (Petsch et al. 1998). The choice of system for endotoxin removal is based on the properties of the bioproducts which needs to be purified. Generally, the interaction between the anionic phosphate in LPS and the cationic ligands on the sorbents are mostly exploited for endotoxin removal (Zhang et al. 2005). The endotoxin removal techniques are also based on hydrophobic interactions between the lipid A part and the sorbent. Endotoxin molecules usually tend to form micelles in aqueous solution (Ritzen et al. 2007). Ultrafilteration can also be employed sue to the difference in sizes of endotoxins and water as well as salt and other small molecules in protein-free solutions. Two-phase extraction systems and affinity chromatography are used based on the physical-chemical interaction between endotoxin and protein to completely remove endotoxin (Anspach and Hilbeck 1995). Detergents can also be used for endotoxin separation from a protein surface; however, there will be the requirement of an additional step to remove surfactant from the product (Ritzen et al. 2007). Generally, the choice of method for endotoxin removal depends upon the rapidness, level of purity, rapidness, cost, difficulty of operation and availabilities of chemicals.

Conclusion

Endotoxin is a form of sugar. It is a structure composed of lipids and sugar complexes. This structure is called a lipopolysaccharide or LPS. This structure is necessary for gram negative bacteria to maintain cell wall integrity. It is a major part of their cell wall and is important for their survival. A systemic inflammatory reaction can occur in conditions where the body is exposed to LPS excessively or systemically (as when small concentrations of LPS enter the blood stream), leading to multiple pathophysiological effects, such as endotoxin shock tissue injury, and death. However, Endotoxin does not act directly against cell or organs but through activation of the immune system, primarily through monocytes and macrophages. Harmful bacteria simulate natural immunity. Endotoxins are believed to be heat stable and some are potent oxidative stress inducers such as superoxide, peroxide, and hypochlorite.

Enormous progress has been made in understanding the molecular mechanisms of LPS. However, the synthesis of diverse endotoxin structures is a complex process. Although different bacteria share similar structural modifications, these modifications may not always have the same influence on pathogenesis. Understanding the biochemistry of lipid, A modifications and their impact on pathogenesis could lead to novel treatment options during Gram-negative bacterial infections.

Bacterial endotoxins' chemical and biological properties have been under investigation for more than a century. During the past two decades, more refined chemical methods for extracting and purifying bacterial endotoxins have been developed. The repeated administration of LPS to man or experimental animals' results in a progressive diminution of some biological effects, especially fever. LPS has been shown to activate the complement system via the alternate pathway.

Furthermore, there is a lack of standard guidelines for the in-vitro evaluation of biomaterials, including the need for regular endotoxin testing and publication of all values along with biomaterial characteristics.

References

Aderem A., Ulevitch R. J., (2000). Toll-like receptors in the induction of the innate immune response. *Nature*; 406: 782–787.

Akira S., Takeda K. (2004). Toll-like receptor signalling. *Nat Rev Immunol*; 4: 499–511.

Aliprantis A. O., Yang R. B., Mark M. R. et al. (1999). Cell activation and apoptosis by bacterial lipoproteins through Toll-like receptor-2. *Science*; 285: 736–739.

Anspach and O. Hilbeck, "Removal of endotoxins by affinity sorbents," *Journal of chromatography A,* vol. 711, no. 1, pp. 81–92, 1995.

Ancuta P., Pedron T., Girard R., Sandstrom G., Chaby R. (1996). Inability of the *Francisella tularensis* Lipopolysaccharide to mimic or to antagonize the induction of cell activation by endotoxins. *Infect. Immun.* ; 64: 2041–2046.

Andrucioli M. C. D., Matsumoto M. A. N., Saraiva M. C. P., Feres M., de Figueiredo L.C., Sorgi C. A., Faccioli L. H., da Silva R. A. B., da Silva L. A. B., Nelson-Filho P. (2018). Successful and failed mini-implants: microbiological evaluation and quantification of bacterial Endotoxin. *J. Appl. Oral Sci.*; 26: e20170631: 1-9. http://dx.doi.org/10.1590/1678-7757-2017-0631.

Babinski K. J., Kanjilal S. J., Raetz C. R. (2002). Accumulation of the Lipid A precursor UDP-2,3-diacylglucosamine in an *Escherichia coli* mutant lacking the *lpxH* gene. *J. Biol. Chem.*; 277: 25947–25956.

Babinski K. J., Ribeiro A. A., Raetz C. R. (2002). The *Escherichia coli* gene encoding the UDP-2,3-diacylglucosamine pyrophosphatase of Lipid A biosynthesis. *J. Biol. Chem.*; 277: 25937–25946.

Beutler, B., Rietschel, E. T., (2003). Innate immune sensing and its roots: the story of Endotoxin. *Nat. Rev. Immunol.*; 3, 169-176.

Bishop R. E. (2005). The Lipid A palmitoyltransferase PagP: molecular mechanisms and role in bacterial pathogenesis. *Mol. Microbiol.*; 57: 900–912.

Clementz T., Raetz C. R. (1991). A gene coding for 3-deoxy-D-*manno*octulosonic- acid transferase in *Escherichia coli*. Identification, mapping, cloning, and sequencing. *J. Biol. Chem.*; 266 (15): 9687-96.

Clementz T., Raetz C. R. (1991). A gene coding for 3-deoxy-D-*manno*octulosonic- acid transferase in *Escherichia coli*. Identification, mapping, cloning, and sequencing. *J. Biol. Chem.*; 266:

Coleman J., Raetz C. R. (1988). First committed step of LipidA biosynthesis in *Escherichia coli*: sequence of the *lpxA* gene. *J. Bacteriol.*; 170: 1268–1274.

Crowell D. N., Anderson M. S., Raetz C. R. (1986). Molecular cloning of the genes for LipidA disaccharide synthase and UDP-*N*acetylglucosamine acyltransferase in *Escherichia coli*. *J. Bacteriol.*; 168: 152–159.

Crowell D. N., Anderson M. S., Raetz C. R. (1986). Molecular cloning of the genes for LipidA disaccharide synthase and UDP-*N*acetylglucosamine acyltransferase in *Escherichia coli*. *J. Bacteriol.*; 168: 152–159.

Darveau R.P., Pham T.T., Lemley K. et al. (2004). *Porphyromonas gingivalis* lipopolysaccharide contains multiple LipidA species that functionally interact with both Toll-like receptors 2 and 4. *Infect. Immun.*; 72: 5041–5051.

Dziarski, R., Gupta, D., (2006). The peptidoglycan recognition proteins (PGRPs). Protein family review. *Genome Biol*. 7, 232.1e232.13.

Galanos C., Luderitz O., Rietschel E. T., *al.* (1985). Synthetic and natural *Escherichia coli* free lipid A express identical endotoxic activities. *Eur J. Biochem.*; 148: 1–5.

Galloway S. M., Raetz C. R., (1990). A mutant of *Escherichia coli* defective in the first step of endotoxin biosynthesis. *J. Biol. Chem.*; 265: 6394–6402.

Garrett T.A., Kadrmas J.L., Raetz C.R. (1997). Identification of the gene encoding the *Escherichia coli* lipid A 4′-kinase. Facile phosphorylation of endotoxin analogues with recombinant LpxK. *J. Biol. Chem.*; 272: 21855–21864.

Garrett T. A., Kadrmas J. L., Raetz C. R. (1997). Identification of the gene encoding the *Escherichia coli* lipid A 4′-kinase. Facile phosphorylation of endotoxin analogues with recombinant LpxK. *J. Biol. Chem.*; 272: 21855–21864.

Girard R., Pedron T., Uematsu S. et al. (2003). Lipopolysaccharides from *Legionella* and *Rhizobium* stimulate mouse bone marrow granulocytes via Toll-like receptor 2. *J. Cell. Sci.*; 116: 293–302.

Golenbock D. T., Hampton R. Y., Qureshi N., Takayama K., Raetz C. R. (1991). Lipid A-like molecules that antagonize the effects of endotoxins on human monocytes. *J Biol Chem*; 266: 19490–19498.

Gutsmann, T.; Schromm, A.; Brandenburg, K. (2007). The Physicochemistry of Endotoxins in Relation to Bioactivity. *Int. J. Med. Microbiol.*, 297, 341–352.

Harald, F. M., (2001). Gleanings of a chemiosmotic eye. *Bioessays*; 23, 848e855.

Heine H., Muller-Loennies S., Brade L., Lindner B., Brade H. (2003). Endotoxic activity and chemical structure of lipopolysaccharides from *Chlamydia trachomatis* serotypes E and L2 and *Chlamydophila psittaci* 6BC. *Eur. J. Biochem.*; 270: 440–450.

Hoffmann, J. A., (2003). The immune response of Drosophila. *Nature* 426, 33e38.

Holtje, J. V., (1998). Growth of the stress-bearing and shape-maintaining murein sacculus of Escherichia coli. Microbiol. *Mol. Biol. Rev.* 62, 181e203.

Hoshino K., Takeuchi O., Kawai T., et al. (1999). Toll-like receptor 4 (TLR4)-deficient mice are hyporesponsive to Lipopolysaccharide: evidence for TLR4 as the LPSgene product. *J Immunol*; 162: 3749–3752.

https://www.horseshoecrab.org/med/endotoxin.html.

Hultmark, D., (1993). Immune reactions in Drosophila and other insects: a model for innate immunity. *Trends Genet*. 9, 178-183.

Jackman J. E., Raetz C. R., Fierke C. A. (1999). UDP-3-*O*-(R-3- hydroxymyristoyl)-*N*-acetylglucosamine deacetylase of *Escherichia coli* is a zinc metalloenzyme. *Biochemistry*; 38: 1902–1911.

Kamio Y., Nikaido H. (1976). Outer membrane of *Salmonella typhimurium*: accessibility of phospholipid head groups to phospholipase C and cyanogen bromide activated dextran in the external medium. *Biochemistry*; 15: 2561–2570.

Kato, Y., Motoi, Y., Taniai, K., (1994). Lipopolysaccharide-hpophorin complex formation in insect hemolymph: a common pathway of lipopolysaccharide detoxification both in insects and in mammals. *Insect Biochem. Mol. Biol.* 24, 547-555.

Kelly T. M., Stachula S. A., Raetz C. R., Anderson M. S. (1993). The *firA* gene of *Escherichia coli* encodes UDP-3-*O*-(R-3-hydroxymyristoyl)- glucosamine *N*-acyltransferase. The third step of endotoxin biosynthesis. *J. Biol. Chem.*; 268: 19866–19874.

Kotani S., Takada H., Tsujimoto M., et al. (1985). Synthetic lipid A with endotoxic and related biological activities comparable to those of a natural lipid A from an *Escherichia coli* Re-mutant. *Infect. Immun.*; 49: 225–237.

Lackie, A. M., (1988). Hemocyte behaviour. *Adv. Insect Physiol.* 21, 85-178.

Leonardo M. R., Silva R. A., Assed S., Nelson-Filho P. (2004). Importance of bacterial Endotoxin (LPS) in endodontics. *J. Appl. Oral Sci.*; 12 (2):93-8.

Loppnow H., Brade H., Durrbaum I. et al. (1989). IL-1 induction-capacity of defined lipopolysaccharide partial structures. *J. Immunol.*; 142: 3229–3238.

Loppnow H., Brade L., Brade H., et al. (1986). Induction of human interleukin 1 by bacterial and synthetic lipid A. *Eur. J. Immunol.*; 16: 1263–1267.

Magalhaes, A. M. Lopes, P. G. Mazzola, C. Rangel-Yagui, T. C. V. Penna, and A. Pessoa, "Methods of endotoxin removal from biological preparations: a review," *Journal of Pharmacy and Pharmaceutical Sciences,* vol. 10, no. 3, pp. 388–404, 2007.

Moran A. P., Lindner B., Walsh E. J. (1997). Structural characterization of the LipidA component of *Helicobacter pylori* rough-and smooth form lipopolysaccharides. *J. Bacteriol.*; 179: 6453–6463.

Morra M., Cassinelli C., Bollati D., Cascardo G., Bellanda M. (2015). Adherent endotoxin on dental implant surfaces: a reappraisal. *J. Oral Implantol.*; 41 (1):10-6.

Nelson-Filho P., Valdez R. M., Andrucioli M. C., Saraiva M. C., Feres M., Sorgi C. A., et al. (2011). Gram-negative periodontal pathogens and bacterial Endotoxin in metallic orthodontic brackets with or without use of an antimicrobial agent: an in vivo study. *Am. J. Orthod. Dentofacial Orthop.*; 140 (6): e281-7.

Nikaido H. (1994). Prevention of drug access to bacterial targets: permeability barriers and active efflux. *Science*; 264: 382–388.

Nishijima M., Bulawa C. E., Raetz C. R. (1981). Two interacting mutations causing temperature-sensitive phosphatidylglycerol synthesis in *Escherichia coli* membranes. *J. Bacteriol.*; 145: 113–121.

Nishijima M., Raetz C. R., (1979). Membrane lipid biogenesis in *Escherichia coli*: identification of genetic loci for phosphatidyl glycerophosphate synthetase and construction of mutants lacking phosphatidylglycerol. *J. Biol. Chem.*; 254: 7837–7844.

Onishi H. R., Pelak B. A., Gerckens L. S., et al. (1996). Antibacterial agents that inhibit LipidA biosynthesis. *Science*; 274: 980–982.

Parillo J. E. (1993). Pathogenic mechanisms of septic shock. *N. Engl. J. Med.*; 328: 1471–1478.

Persing D. H., Coler R. N., Lacy M. J. et al. (2002). Taking toll: LipidA mimetics as adjuvants and immunomodulators. *Trends Microbiol.*; 10: S32–S37.

Petsch D, E. Rantze, and F. B. Anspach, "Selective adsorption of endotoxin inside a polycationic network of flat-sheet microfiltration membranes," *Journal of Molecular Recognition,* vol. 11, no. 1–6, pp. 222–230, 1998.

Pilione M. R., Pishko E. J., Preston A., Maskell D. J., Harvill E. T. (2004). PagP is required for resistance to antibody-mediated complement lysis during *Bordetella bronchiseptica* respiratory infection. *Infect. Immun.*; 72: 2837–2842.

Preston A., Maxim E., Toland E., et al. (2003). *Bordetella bronchiseptica* PagP is a Bvg-regulated lipid A palmitoyl transferase that is required for persistent colonization of the mouse respiratory tract. *Mol. Microbiol.*; 48: 725–736.

Raetz C. R. (1996). In: Niedhardt F (ed.) *Escherichia coli and Salmonella Cellular and Molecular Biology*. Washington, DC: *Am. Soc. Microbiol.*; 1035–1063.

Raetz C. R., Purcell S., Meyer M. V., Qureshi N., Takayama K. (1985). Isolation and characterization of eight LipidA precursors from a 3-deoxy-D-*manno*-octylosonic acid-deficient mutant of *Salmonella typhimurium. J. Biol. Chem.*; 260: 16080–16088.

Raetz C. R., Roderick S. L. (1995). A left-handed parallel beta helix in the structure of UDP-*N*-acetylglucosamine acyltransferase. *Science*; 270: 997–1000.

Raetz C. R., Whitfield C. (2002). Lipopolysaccharide endotoxins. *Annu. Rev. Biochem.*; 71: 635–700.

Ramachandran G. (2014) Gram-positive and gram-negative bacterial toxins in sepsis: A brief review. *Virulence*. 5 (1): 213–218.

Ray B. L., Raetz C. R. (1987). The biosynthesis of Gram-negative Endotoxin. A novel kinase in *Escherichia coli* membranes that incorporates the 4′-phosphate of lipid A. *J. Biol. Chem.*; 262: 1122–1128.

Ritzen, J. Rotticci-Mulder, P. Stromberg, and S. R. Schmidt, "Endotoxin reduction in monoclonal antibody preparations using arginine," *Journal of Chromatography B,* vol. 856, no. 1-2, pp. 343–347, 2007.

Rietschel E. T., Kirikae T., Schade F. U., et al. (1994). Bacterial Endotoxin: molecular relationships of structure to activity and function. *FASEB J.*; 8: 217–225.

Rietschel E. T., Westphal O. (1999). History of endotoxins. In: Brade H. (ed) *Endotoxins in Health and Disease*. New York: Marcel Dekker; 1–30.

Robey M., O'Connell W., Cianciotto N. P. (2001). Identification of *Legionella pneumophila rcp*, a *pagP*-like gene that confers resistance to cationic antimicrobial peptides and promotes intracellular infection. *Infect. Immun.*; 69: 4276–4286.

Sandstrom G., Sjostedt A., Johansson T., Kuoppa K., Williams J. C. (1992). Immunogenicity and toxicity of Lipopolysaccharide from *Francisella tularensis* LVS. *FEMS Microbiol. Immunol.*; 5: 201–210.

Schnaitman C. A., Klena J. D., (1993). Genetics of lipopolysaccharide biosynthesis in enteric bacteria. *Microbiol. Rev.*; 57: 655–682.

Shimazu R., Akashi S., Ogata H. et al. (1999). MD-2, a molecule that confers lipopolysaccharide responsiveness on Toll-like receptor 4. *J Exp Med*; 189: 1777–1782.

Suda Y., Kim Y. M., Ogawa T. et al. (2001). Chemical structure and biological activity of a lipid A component from *Helicobacter pylori* strain 206. *J. Endotoxin Res.*; 7: 95–104.

Takayama K., Qureshi N., Mascagni P., Nashed M. A., Anderson L., Raetz C. R. (1983). Fatty acyl derivatives of glucosamine 1-phosphate in *Escherichia coli* and their relation to LipidA. Complete structure of A diacyl GlcN-1-P found in a phosphatidylglycerol-deficient mutant. *J. Biol. Chem.*; 258: 7379–7385.

Takayama K., Qureshi N., Mascagni P., Nashed M. A., Anderson L., Raetz C. R. (1983). Fatty acyl derivatives of glucosamine 1-phosphate in *Escherichia coli* and their relation to LipidA. Complete structure of A diacyl GlcN-1-P found in a phosphatidylglycerol-deficient mutant. *J. Biol. Chem.*; 258: 7379–7385.

Tobias P. S., Soldau K., Ulevitch R. J. (1986). Isolation of a lipopolysaccharide-binding acute phase reactant from rabbit serum. *J. Exp. Med.*; 164: 777–793.

Trent M. S. (2004). Biosynthesis, transport, and modification of lipid A. *Biochem. Cell Biol.*; 82: 71–86.

Ulevitch R. J., Tobias P. S., (1999). Recognition of gram-negative bacteria and Endotoxin by the innate immune system. *Curr. Opin. Immunol.*; 11: 19–22.

Vaara M. (1993). Antibiotic-supersusceptible mutants of *Escherichia coli* and *Salmonella typhimurium*. *Antimicrob. Agents Chemother.*; 37: 2255–2260.

van Deuren M., Brandtzaeg P., van der Meer J. W. (2000). Update on meningococcal disease with emphasis on pathogenesis and clinical management. *Clin. Microbiol. Rev.*; 13: 144–166.

Vishnupriya S. (2018). Bacterial endotoxin-lipopolysaccharide; structure, function, and its role in immunity in vertebrates and invertebrates. *Agriculture and Natural Resources* 52: 115e120.

Vogel S. N., Madonna G. S., Wahl L. M., Rick P. D. (1984). *In vitro* stimulation of C3H/HeJ spleen cells and macrophages by a lipid A precursor molecule derived from *Salmonella typhimurium*. *J. Immunol.*; 132: 347–353.

Wenqiong S., Xianting D. (2015). Methods of Endotoxin Detection. *Journal of Laboratory Automation*. 20 (4): 354–364.

Whitfield C. (1995). Biosynthesis of lipopolysaccharide O antigens. *Trends Microbiol.*; 3: 178–185.

Williamson J. M., Anderson M. S., Raetz C. R. (1991). Acyl-acyl carrier protein specificity of UDP-Glc NAc acyltransferases from gram negative bacteria: relationship to LipidA structure. *J. Bacteriol.*; 173: 3591–3596.

Young K., Silver L. L., Bramhill D. et al. (1995). The *envA* permeability/cell division gene of *Escherichia coli* encodes the second enzyme of LipidA biosynthesis. UDP-3-*O*-(R-3-hydroxymyristoyl)-*N*acetylglucosamine deacetylase. *J. Biol. Chem.*; 270: 30384–30391.

Zhang J. P., Q. Wang, T. R. Smith, W. E. Hurst, and T. Sulpizio, "Endotoxin removal using a synthetic adsorbent of crystalline calcium silicate hydrate," *Biotechnology Progress,* vol. 21, no. 4, pp. 1220–1225, 2005.

Chapter 2

Extraction and Characterization of Endotoxins

Srishti Prashar* and Prakriti Sharma
PhD Scholar, College of Animal Biotechnology,
Guru Angad Dev Veterinary and Animal Sciences University, Ludhiana, Punjab, India

Abstract

Endotoxins or Lipopolysaccharide are present as the major surface component of the outer membrane of Gram-negative bacteria and are responsible for activation of the innate immune system. LPS is made up of lipid A, core oligosaccharides and O Antigen repeats where the lipid A domain of LPS is known as endotoxin. It anchors the molecule in the outer membrane and is the bioactive component recognized by TLR4 during infection. The host response to the presence of endotoxin is dependent on both the severity of the infection and the lipid A/ endotoxin structure of the invading microbe. So, isolation and examination of biologically active cellular components of virulent bacteria are required however, due to their specific structure and amphipathic property, analysis of LPS is difficult. Therefore, this chapter explores the different approaches for the extraction as well as physical and chemical characterization of endotoxins/LPS.

Keywords: Endotoxin, LPS, Lipid A, Extraction, Purification, Characterization

* Corresponding Author's Email: prasharsrishti01@gmail.com.

In: Endotoxins and their Importance
Editors: Arif Pandit and R. S. Sethi
ISBN: 978-1-68507-839-3

Introduction

The enhancement in infectious disease is a worldwide issue and emerging pathogens that are resistant to drugs are becoming a continuing risk to both public health and agriculture (Stromberg et al., 2017). Accurate and rapid identification along with characterization of pathogens is crucial for the implementation of preventative measures to limit this problem. Lipopolysaccharides (LPS) are molecules that have both hydrophilic and hydrophobic regions and are considered as a key component associated with the outer cell membrane of Gram-negative bacteria (Kumar et al., 2011, Suand & Ding, 2015). Further, it becomes an ideal component for detection and diagnostics due to its various properties such as an active infection indicator, specificity for serogroup and better stability than its protein counterparts (Munford, 2008). Hence, it is preferred for pathotypes identification and characterization for the timely treatment of infections (Jauho et al., 2000).

Various Gram-negative bacteria including *Acinetobacter* (Joly-Guillou, 2005), *E.coli* (Leimbachet al., 2013), *Helicobacter* (Yamaoka & Graham, 2014), *Hemophilus* (Livorsi et al., 2012), *Pseudomonas* (Mahajan-Miklos et al., 1999), *Salmonella* (Mather et al., 2013), *Shigella* (Caboni et al., 2015), *Yersinia* (Fàbrega & Vila, 2012) and Enterobacteriaceae family that contaminate food, water, and soil that results in nosocomial infections (Deisingh & Thompson, 2004), liver damage (Magdaleno et al., 2017), neurological degeneration (Gomes et al., 2017), lung disorders (Garcia et al., 2012). LPS structure and function play an important role in the identification of *E. coli* serogroup, in vaccine design and in the therapeutic interventions due to which it is also known as a virulence factor that releases in the environment by Gram-negative bacteria (Stromberg et al., 2017). Hence it becomes essential to detect and characterize the endotoxins for epidemiology, disease control, and treatment purposes.

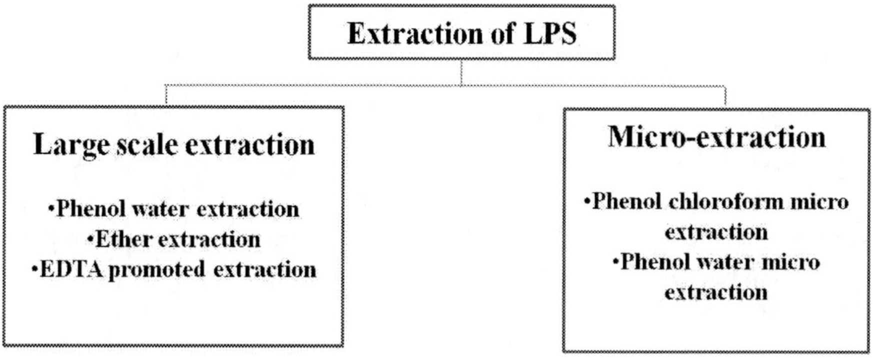

Figure 2.1. Methods used for extraction of LPS.

2.1. LPS Structure

LPS is involved in host cell attachment and initiates the infection along with protection from complement-mediated killing. LPS consists of lipid A, core oligosaccharide, and O-antigen repeats (Farhana & Khan, 2021). Among the various components of LPS, lipid A is more conserved in structure as compared to others. The lipid A actively participates in the endotoxin activity of the molecules but without harming any tissue or the white blood cells, hence is an indicator for bacterial presence (Bertani & Ruiz, 2018). Therefore, LPS visualization is necessary as it is helpful in the identification of strain lineages and mutant characterization (Davis et al., 2012). Based on size and structures, LPS can be divided into smooth type LPS (S-LPS) and rough type LPS (R-LPS). The S-LPS has different O-antigen repeats whereas R-LPS varies in core oligosaccharides and due to these structural variations various techniques have been evolved for the extraction, purification, and analysis of LPS (Wang et al., 2010).

2.2. Extraction of LPS

2.2.1 Large Scale Extraction

Single method is not enough for LPS extraction on large scale due to its variable structure and different amphipathic properties. Various methods such as Phenol water extraction, Ether extraction and EDTA promoted extraction are preferred for specific LPS groups.

2.2.1.1. Phenol–Water Extraction

Phenol and water mixture (45:50, v/v) method separate LPS and proteins into two different phases i.e., phenol and aqueous phase (Sarmikasoglou and Faciola, 2021, Wang et al., 2010). Proteins are present in the phenol phase whereas, nucleic acids along with polysaccharides are found in the aqueous phase (Davis & Goldberg, 2012). A total of 100 ml bacterial culture was pasteurized at 100°C for 2 hours followed by cooling at 28°C. Add equal amount of phenol in the bacterial suspension followed by centrifugation at 18,000g for 15 minutes to get three layers i.e., aqueous, phenol, and the interface. Aqueous layer is removed carefully followed by double washing of pellet with sterilized PBS (pH 7.2). The pellet is then suspended by vortexing in 15 ml of Tris-Cl buffer with 2% Sodium Dodecyl Sulphate, 4% 2-mercaptoethanol, and 2mM $MgCl_2$.The dissolved pellet is further incubated at 65°C for solubilization of bacteria.

There may be possibility of protein contamination, specially lipoproteins in LPS so proteinase hydrolysis and other purification processes are required (Apicella, 2008). For this, the solution will be treated with 1ml Proteinase K (@100 μg/ml followed by overnight incubation at 37°C. After incubation, 2ml of 3 M sodium acetate was used to enhance the LPS precipitation by ethanol followed by overnight incubation at -20°C. Then, centrifugate at 4000 g for 15 mins and discard the supernatant. The remaining precipitates are dissolved in the mixture of sterilized distilled water and 3M sodium acetate (9:1) by vortexing. Further, add 20ml of cold absolute ethanol followed by vortexing the mixture again and incubating the suspension overnight at -20°C for precipitation to remove the SDS residue from LPS. After the final centrifugation, suspend the precipitate in 9ml of 10mM Tris-Cl, DNase I (100 μg/mL), and RNase (25 μg/mL). After incubation for 4 hours at 37°C, assure the removal of all residual DNA/RNA contamination. Finally, to remove residual protein contamination place the LPS mixture in the water bath at 65°C for a half-hour followed by addition of equal volume of 90% phenol and incubate for 15 mins at 65°C, then, quickly cool the mixture and centrifuge at 6000g for 15 min to remove the aqueous layer and to extract the phenol layer. Again, after centrifugation the achieved aqueous layer was mixed with the first aqueous extract and dialyzed by using triple distilled water and the bottom phenol layer was discarded. The LPS then lyophilised to achieve the powder form which is stored at -20°C for further use (Kalambhe et al., 2017, Apicella, 2008).

2.2.1.2. Ether Extraction

This method was opted for R-LPS extraction to overcome the limitation of the phenol–water extraction method that is a frequent partition of R-LPS into the phenol phase. A monophasic solution consisting of liquid phenol, chloroform, and petroleum ether (PCP) is utilized during this method. LPS is solubilized completely whereas, S-LPS, proteins, nucleic acids, and carbohydrates, and other polysaccharides are insoluble. The cells were dried after harvesting and washed with distilled water. The dried bacteria is placed in 200mL PCP (2:5:8) followed by homogenization and centrifugation at 5000g. Filter the supernatant in a flask and re-extraction of the pellet with a PCP (2:5:8), and continue it with centrifugation as above. Pooled the supernatant into the first extract and remove the petroleum ether and chloroform in the extracts by a rotary evaporator followed by centrifugation to achieve the precipitated LPS by ether washing and drying in vacuum. The precipitated LPS dissolved in distilled water followed by warming and shaking to achieve a viscous solution. At last, centrifuge to get LPS pellets dissolved in water and dried (Wang et al., 2010).

2.2.1.3 EDTA Promoted Extraction

This method can be used for both S-LPS and R-LPS due to the size diversification of LPS molecules (Wang & Quinn, 2010). In this, SDS and EDTA are preferred for the precipitation of contaminants including murein or polymer and proteins followed by drying of bacterial cells and suspension in the solution having Tris–HCl, MgC12, pancreatic DNase I and RNase A. Break the cells and add DNase I and RNase A to final concentrations followed by incubation of the suspension at 37°C. Then, add EDTA, SDS, and Tris–HCl (pH 8.0) to the final volume after digestion followed by vortex and centrifugation of the solution for the removal of peptidoglycan. Pronase is added to the supernatant for the digestion of the dead cells and incubate overnight followed by the adding double volume of $MgCl_2$ dissolved in 95% ethanol. Cool the mixture, centrifuge and resuspend the pellets in Tris–HCl, and EDTA followed by sonication. The mixture should be incubated after setting pH to denature the SDS-resistant proteins present on the outer membrane. Cool the solution and set the pH to 9.5 follow up by overnight incubation at 37°C with constant stirring. Centrifuge the solution after precipitating LPS and then the achieved pellet will be resuspended in Tris–

HCl followed by sonication, and centrifugation for the removal of Mg^{2+}-EDTA crystals and repeat it to achieve the LPS pellet (Subhi et al., 2017).

2.2.2 Micro-Extraction of Lipopolysaccharides

This extraction method is used when small bacterial concentrations are available and small LPS concentration is required. The LPS extracted by these methods may not be with high purity but can be utilized for analysis. In this, lysing buffer was used for resuspension of cells to get harvested cells and incubation can be done at 100°C. Then, proteinase K is added followed by incubation at 60°C and the detection of LPS can be done by using gel electrophoresis or blotting techniques (Wang et al., 2010).

2.2.2.1. Phenol–Water Micro-Extraction

This is a rapid method in which 2 × 109 colonies forming unit (CFU) of bacteria are washed in phosphate-buffered saline containing $CaCl_2$ and $MgCl_2$ and resuspended in distilled water. Hot phenol is added with equal volume followed by vigorous agitation at 68°C. The mixture is cooled and the phenol–water phases are separated by centrifugation. Remove the aqueous phase and add distilled water to the phenol phase, and repeat the extraction. Add sodium acetate to the final concentration after pooling of the aqueous phase followed by the addition of ethanol. After overnight incubation, collect the insoluble crude LPS by centrifugation and re-dissolved it in distilled water, repeating the precipitation. Then, again dissolved the final product in distilled water and stored at –20°C (Wang et al., 2010).

2.2.2.2. Phenol–Chloroform Micro-Extraction

Phenol is usually used along with the combination of chloroform to confirm the proper separation between aqueous and organic phases. In this, an overnight bacterial cells culture is centrifuged and the pellet is suspended in triethylamine (TAE) buffer. The suspended pellet is then mixed with alkaline solutions which includes SDS, Trizma base, and 2 M NaOH dissolved in H_2O. The mixture is incubated at 60°C for approximately an hour and then mixed with phenol-chloroform (1:1, v/v). Centrifuge the mixture and precipitate by adding ethanol which is then dissolved in Tris hydrochloride (pH 8.0)-100 mM sodium acetate and repeat the precipitation. Dissolve the final precipitates of LPS in H_2O and visualize them by using ethidium bromide (EtBr) within 30 min (Kido et al., 1990).

2.3. Isolation of Lipid A

The classical method is widely preferred due to the insolubility of lipid A in water, it forms a precipitate that can be extracted by centrifugation. Another method used to extract the small amounts directly from whole cells without a prior LPS extraction. The third method uses relatively mild hydrolysis for the preservation of acid-labile phosphorylation sites and lipid A backbone head groups. This isolation method is specifically utilized for lipid A isolation to analyze its structure.

2.3.1. Acetic Acid Hydrolysis Method

During this method, LPS is solubilized in aqueous triethylamine followed by adding acetic acid to a final concentration of 1.5% (v/v). Heat the mixture for 2 hours and then cool. Quantitatively precipitate lipid A by adding HCl to a final pH of 1.5 and centrifuge lipid A which is not soluble at 2000*g* with subsequently 2 times washing with distilled water followed by freeze-drying (Wang et al., 2010).

2.3.2 Microlipid A Extraction from Whole Cells

The lyophilized freshly washed bacterial cells were suspended in isobutyric acid–ammonium hydroxide and kept at 100°C for 2 h in a screw-cap test tube under magnetic stirring. Quickly cool the mixture followed by centrifugation at 2000*g* for 15 min. Dilute the supernatant in water and lyophilize it by double washing of sample with methanol and then repeat the centrifugation on the same conditions. Then finally solubilize the insoluble lipid A and extract in a mixture of Chloroform-Methanol-Water (El Hamidi et al., 2005).

2.3.3 Lipid A Isolation (For Structural Analysis)

Dissolve LPS in sodium acetate (pH-4.5) which contains 1% SDS and place in sonicator to dissolve the sample. Then, heat the sample and dry by using Speed Vac continued with the washing of sample with distilled water for the removal of SDS and acidified followed by centrifugation. Again, the sample is washed with non-acidified ethanol which is then centrifuged and lyophilized to yield fluffy white solid lipid A (Wang et al., 2010).

2.4. Purification of Lipopolysaccharides

There may be possibilities of contamination in LPS which is directly extracted from bacteria of other macro molecules such as nucleic acids, peptidoglycan, phospholipids, lipoproteins, and others. Although contaminated LPS also results in confounded cellular responses. For example, a similar response has been observed from LPS and peptidoglycan to bacterial infection through Toll-like receptors 4 and 2 (Girardin et al., 2003). Therefore, it is required to remove the contaminants from the LPS preparations and for these various methods have been accepted to dissociate the contaminant aggregates of LPS and its purification. The effectiveness of these methods relies on the physicochemical interactions that occur between LPS and the contaminants (Stephens et al., 2021).

Phospholipids are considered as the crucial contaminants during starting phase of LPS extraction as it is one of the major membrane components, whereas the level of phospholipids contaminants varies which can be identified by several techniques including thin-layer chromatography (TLC) or mass spectrometry. Treatment of chloroform-methanol mixture is preferred for the removal of contaminated phospholipids due to their solubility in the solvent, but LPS remains separated. Whereas UV absorbance with proper wavelength was preferred to measure the contamination degree of lipoproteins which are another LPS samples contaminants and proteinase K is utilized for the degradation of Lipoproteins. Peptidoglycans are also LPS contaminants responsible for activating the same host response as LPS, hence needed to be eliminated from the sample. SDS and mineral acid are used for the peptidoglycan removal and its dissociation (Hirschfeld et al., 2000).

2.4.1. Purification Methods

2.4.1.1. Gel Filtration Chromatography of LPS

This technique is widely used to isolate heterogeneous LPS isolated from bacteria, especially for micro LPS isolation without having lipid A moiety. The polysaccharide in lipid A will be released by complete spitting of acid-labile Kdo linkage followed by removal of fatty acids connected with esters, reducing non-polar linkage between LPS components and facilitating their separation. For this several Sephadex columns including Sephadex, Sepharose 4B is widely used for intact or truncated LPS separation. The column selection relies on the repeating arrangement of O-antigen in LPS. In this, materials having detergent are used for the separation and elution of intact LPS. Moreover, elution buffer with detergents including SDS and others are used for LPS molecule disaggregation. The multivalent cations are removed from the sample by using EDTA along with monovalent cations in column and dialysis buffers (De Haas et al., 2000, Wang et al., 2010).

2.4.1.2. Ion-Exchange Chromatography of LPS

LPS has negatively charged lipid A along with oligosaccharide regions due to which it can be cleaved by anion-exchange chromatography and further can be analyzed by several techniques including mass spectrometry and NMR (Lukasiewicz et al., 2006, Vilches et al., 2007, Kutschera et al., 2021). This method is also efficient for the purification of less hydrophobic LPS. For the reduction of hydrophobicity, the hydrolyzed LPS results in the removal of lipid A fatty acyl chains.

2.4.1.3. Capillary Electrophoresis (CE) of LPS

CE has improved heat dissolution if compared with gel electrophoresis (Annamária, 2012). This is the separation technique with high resolution and is usually preferred for analysis of mixtures of low-molecular weight molecules including amino acids, carbohydrates, and nucleotides (DeLaney et al., 2019). Along with this, it is used to purify the LPS as it has groups with high polarity, phosphoethanolamine, carboxylic acid, and essential or non-essential amino acids. Whereas, the glycoform populations in LPS were easily identified by using the combination of CE to electrospray and mass spectrometry (CE-ES-MS) (Li et al., 2004, Li et al., 2005, Li et al., 2007). Whereas, species in heterogeneous native R-LPS were identified or detected by usingFourier-transform ion cyclotron resonance mass spectrometer (FT-ICR MS) in combination with CE due to its property of providing high resolution and accurate mass of LPS (Hubner & Lindner, 2009).

2.4.2. Lipid A Purification

DEAE-cellulose ion-exchange chromatography is preferred for Lipid A purification, which relies on the phosphate and pyrophosphate groups present in the lipid A moiety (El Hamidi et al., 2005). Along with this, lipid A can be isolated and detected by TLC as well (Zhou et al., 1999). In addition, chromatography techniques are also used for the isolation of lipid A by depending on the molecular polarity disparity. For example, the use of Bio-Sil to separate the components of lipid A from *Rhizobium etli* (Que et al., 2000) Whereas a C18 reverse column was utilized to separate crude lipid A from *Rhizobium trifolii* (Jeyaretnam, 1998). Moreover, various protocols have been successfully implanted to separate monophosphoryl lipid A (Qureshi et al., 1997). Briefly, the conversion of monophosphoryl lipid A into the free acid can be done by passing monophosphoryl lipid A through Chelex 100 (Na+) and Dowex 50 (H+) columns followed by methylation with diazomethane. Moreover, the HPLC technique can also be used to isolate monophosphoryl lipid A (Qureshi et al., 1997).

2.4.2.1. Micro-purification Method

This method is raised from nucleic acids and polypeptides micro-purification with zinc-imidazole staining which is negatively charged in SDS-polyacrylamide gels and is dependent on LPS elution from polyacrylamide gels (Hardy et al., 1998, Pupo et al., 2000). As zinc-imidazation detection does not influence the structure and biological characters so, SDS-polyacrylamide gels are preferred to purify and recover the LPS by passive extraction (Pupo et al., 2000). Moreover, about 70–80% rough and semi-smooth LPS can be retrieved after 3 h of extraction by using average size gel microparticles in water (Hardy et al., 1998). But only 5 to 10% of smooth LPS type is retrieved because of high aggregation in gel, the low water solubility, and diffusion rate. To enhance the smooth LPS reproducibility, distilled water can be replaced by 1% SDS or DOC or TEA. However, 5% TEA is preferred among them due to its easily removable property by simple evaporation (Pupo et al., 2000).

2.5. Analysis of Lipopolysaccharides

Identification or analysis of mutated strains of key genes involved in LPS biosynthesis is necessary to acknowledge the LPS role in the pathogenesis of bacteria. Therefore, there is the requirement of a quick and sensitive method for screening that can be able to differentiate between the variation in the structural composition of LPS. For this, sodium dodecyl sulfate-polyacrylamide gel electrophoresis (SDS-PAGE), followed by silver staining is a widely preferred analytical method (Fomsgaard et al., 1990). Sensitive reverse staining with zinc-imidazole salts is developed to overcome the cons of silver staining, which include the irreversible fixation and chemical modification (Hardy et al., 1997, Hardy et al., 1998).

2.5.1. Electrophoresis of LPS

SDS-PAGE is used to observe the structural diverseness of LPS by assessing the profile of the altered electrophoretic band (Wang et al., 2010). Tricine has been utilized instead of glycine in both gel and buffer used for electrophoresis to obtain a better resolution of LPS (Lesse et al., 1990). To eliminate the LPS aggregation, SDS can be replaced by sodium DOC during electrophoresis (Komuro & Galanos, 1988). The minor variation in the low molecular weight band of LPS in polyacrylamide gels has been observed by bilayer stacking gel (Inzana & Apicella, 1999).

2.5.2. Staining Methods

2.5.2.1. Silver Stain

This technique is widely used to identify the proteins in gels as it binds to side chains of amino acids. Moreover, in silver stain, the polycarbohydrate part in the LPS molecule is considered as the sensitive part as oxidation of hexoses confirms the availability of aldehyde groups for the next reaction occurs with the silver nitrate. However, fatty acids number in lipid A regulates the LPS fraction retention in the SDS-PAGE during starting fixation and oxidization steps. Improvement in modified staining method was done by deleting the fixing step and increasing the staining oxidation time to overcome the cons of the technique to not to detect the LPS fractions with deacylated S-LPS (Fomsgard et al., 1990). By this method less amount of LPS is sufficient to visualize in polyacrylamide gels.

2.5.2.2. Ethidium Bromide (EB) Staining

This method was developed to overcome the limitations of the silver staining method and is appropriate to stain the LPS which possess acidic O-specific polysaccharides. LPS which contains long O-specific polysaccharide chains with high molecular weight is stained by EB in a better way as compared to low-molecular-weight LPS. During this process, the gel is removed from the glass plate after electrophoresis, and immerses it in an EtBr solution for 10 seconds. Put the gel in distilled water for destaining for a half hour and a transilluminator is used to observe the bands at the wavelength of 302 nm (Wang et al., 2010).

2.5.2.3. Zinc-Imidazole Stain

Disadvantages of silver staining includes high background that results in reduced sensitivity, time-consumable, toxicity, high priced, irreversible fixation and LPS chemical modification. To overcome these drawbacks, a sensitive reverse staining method has emerged which uses zinc and imidazole salts to recover the LPS from gel to analyze the structure (Hardy et al., 1998). Imidazole is used to develop the background for staining and leaves the complex of zinc-LPS with transparent bands (Wang et al., 2010).

2.6. Characterization of LPS

Different analytical methods used to characterize the chemical structure of lipid A species isolated from whole cells are NMR, TLC, MS-based analysis, and others (Henderson et al., 2013).

2.6.1. Nuclear Magnetic Resonance (NMR)

NMR allows elucidation of non-destructive structural as well as provides structural details associated with glycosidic linkages, indubitable positions assignment of acyl chain, and sites for attachment of lipid A alteration like a phosphoethanolamine (Wang et al., 2006).

2.6.2. Immunoblotting Method

The immunoblotting technique is used to characterize the LPS immunochemistry due to the antibodies binding property of the O-antigen in LPS. The specific O-antigen binds with the monoclonal antibodies which act as primary antibodies and visualized by using a secondary antibody which is alkaline phosphatase-conjugated (D'Haeze et al., 2007; Yokota et al., 2000). So, antiserum is used as a primary antibody, followed secondary antibody IgG conjugated with peroxidase which is against the primary antibody. The soaked LPS-containing gel from the transfer buffer shifted to a nitrocellulose membrane after electrophoresis. The blots are performed serially followed by primary antibody, secondary antibody and the bound antibody further can be detected by using color-generating agents (Wang et al., 2010).

2.6.3. Thin-Layer Chromatography (TLC)

TLC method is preferred for fast or quick analysis but cannot provide proper information related to the fine chemical structure. This method separates lipid A species which is colored with the chromogenic agent after chromatography. The dissolved lipids in $CHCl_3$/MeOH are spotted onto a TLC plate i.e., Silica Gel 60 and developed in the solvent $CHCl_3/MeOH/H_2O/NH_4OH$. The lipids are then visualized after drying by discoloring or burning at 145°C (Que et al., 2000). Lipid A can be analyzed by autoradiography techniques due to its detection sensitivity and also by ESI MS (Wang et al., 2006)

2.6.4. Mass Spectrometry (MS)

MS-based protocols including multi-stage MSn strategies have become an indispensable tool used to characterize the structure of lipid A. Moreover, photodissociation mass spectrometry results in activation and fragmentation of ions using photons and detect productions by using mass spectrometry. This combination has become an affordable tool for biological molecules characterization. Structural, binding energies, isomerization along conformation information about ions rely on the dissociation of ions in the gas phase. For this, classic collisional-based methods are often implemented among all other methods. Collisionally activated dissociation method (CAD) also known as collision-induced dissociation (CID) is a vital part of every commercial tandem mass spectrometer (Brodbelt, 2014). The low-energy CID has been widely employed to identify and characterize various lipid A structures from various bacterial species with the help of ultraviolet photodissociation (UVPD) and infrared multiphoton dissociation (IRMPD). The Lipid A fragmentation can also be studied by using activated electron photo detachment (a-EPD) method. a-EPD method uses 193 nm photons that induce the charged reduced radicals which are eventually separated by collisional activation (Madsen et al., 2011).

However, CID, IRMPD can differentiate the products selectively at different stages of phosphorylation due to the enhanced photo absorption cross-sections and thus dissociate the phosphate-containing species. High yield can be attained by both UVPD and a-EPD as compared to CID- IRMPD due to single-photon absorption as well as increased energy. UVPD (193 nm) increase the division between the amine and carbonyl groups on the 2- and 2'-linked primary acyl chains (Madsen et al., 2011).

2.7. Characterization of Different Bacteria

2.7.1. Francisella Species

4 species of genus Francisellatularensis are tularensis (Type A), holartica (Type B), mediasiatica, and novicida. The structure of lipid A isolated from F.tularensis and F. novicida (U112) were detected using mass spectrometry, gas chromatography, nuclear magnetic resolution (NMR), and various other chemical methods (Phillips et al., 2004, Vinogradov et al., 2002). The elucidated primary structure showed that lipid A in all strains are similar having β-(1,6)-linked glucosamine disaccharide with fatty acids linked with amide at the 2 (3-OH) Carbon 18 and 2 (3-OH C18-O-C16) positions and fatty acids linked with ester at 3 (3-OH C18). Furthermore, the base structure of lipid A in *F. holartica* and *F novicida* comprise a phosphate moiety at the 1 position of the glucosamine residue which wasreducedand further substituted with the positively charged sugar, galactosamine, and α-linked in *F novicida* (Vinogradov et al., 2002). Francisellanovicida U112 phospholipids consist of various phospholipids including phosphatidylethanolamine, phosphatidylcholine, and two species of lipids A which comprises 15% of the total phospholipids and less than 5% are covalently associated. Both glucosamine disaccharides species, are primary 3-hydroxystearoyl chains acylated on secondary palmitoyl residue whereas, minor isobaric species have a 3-hydroxypalmitoyl chain instead of 3-hydroxystearate (Wang et al., 2006, Schilling et al., 2007). Along with this, the central region of *F holartica* strain (LVS) and *F.novicida* (U112) linked to lipid A through eight-carbon sugar, 3-deoxy-D-mannooctulosonic acid (Kdo)(Vinogradov& Perry,2004). The only difference between the F. novicida and F. holartica is the presence of an extra residue of α-glucose which is bound with the β-Glc to the central inner core α-Man residue. Whereas, the central region of both *F holartica* strain (LVS) and *F.novicida* (U112)lacks phosphate modifications and includes a single Kdo residue, which indicates the presence of a Kdo-specific hydrolase enzyme (Stead et al., 2005). O antigen of F. tularensis and F. holartica has the same

structure whereas O-antigen of *F. novicida* shares two same internal carbohydrate residues (α-D-GalNAcAN- α-D-GalNAcAN), but different for the outer two residues (α-D-GalNAcAN and β-DQui2NAc4NAc for *F. novicida* versus β-DQui4NFm and β-D-QuiNAc for F. tularensis and F. holartica) (Prior et al., 2003, Thomas et al., 2007).

2.7.2. Yersinia Pestis

The *Yersinia* genus of the Enterobacteriaceae family has two enteropathogenic species, *Y.pseudotuberculosis* and *Y. enterocolitis*, involve in chronic intestinal infections. *Y. pestis* is a pathogenic bacterium that forms rough-colony and thereby produces R-type LPS with its carbohydrate moiety which is bound to an oligosaccharide called the core, an intermediate region between the O-antigen and lipid A, whereas, the Smooth-type LPS form smooth-colony-forming which also has a polysaccharide chain (O-antigen) with oligosaccharide repeating units. The LPS is made up of a small carbohydrate (oligosaccharide) chain which is linked with lipid A with 1,4'-phosphorylated glucosamine double residues which are acylated with 3-hydroxymyristate along with four residues and are called primary acyl groups. Along with this, the glucosamine residues also consist of D-glucose, L-glycero-D-mannoheptose, acetyl, lipid A, D-glycero-D-mannoheptoseB-hydroxymyristate, glucosamine, 3-deoxyoctulosonic acid, phosphate, and protein, whereas other secondary acyl residues i.e., laurate and palmitoleate are connected with primary fatty acids hydroxyl groups in the residue of glucosamine and carry the central oligosaccharide (Rebeil et al., 2004, Rebeil et al., 2006). The replacement of glucosamine moiety of lipid A can be done with *Y. pestis* with phosphate, amide-linked and also by β-hydroxymyristate. The absence of significant amounts of additional fatty acids affects the complexity of the structure of lipid A if compare with gram-negative bacteria. The outer oligosaccharide region is absent in *Y. pestis* and components of monosaccharides of its inner region are the representative of the *Yersinia* species. Therefore, the heptose residue presents at distance from lipid A (LD-HepIII) along with a residue of *D-glycero-D-manno*-heptose. Furthermore, *N*-acetyl-*D*-glucosamine, present in non-stoichiometric amounts, replaces the central heptose residue (LD-HepII) (Knirel et al., 2005). "Core" region of LPS from the Enterobacteriaceae consists of maximum sugar, except for the D-glycero-D-mannoheptose. The estimated molecular mass of the aggregated LPS is 1.6 x 108. Moreover, acylation of lipid A relies on environmental conditions such as temperature and different forms of tetradecyl, pentadactyl, and hexadecyl whereas the triacyl form is a common one. Elevated temperature decreases the lipid A acylation (Knirel et al., 2005, Dentovskaya et al., 2008, Knirel et al., 2008). Residues of cationic monosaccharide, 4-amino-4-deoxy-*L* arabinose (Ara4N) cause glycosylation of the lipid A phosphate groups. At low temperature, the glycosylation of both phosphate groups of LPS form is stoichiometric, whereas high temperature reduces Ara4N (Knirel et al., 2012) with the insertion of additionally phosphorylated phosphate leading to the formation of a diphosphate group (Dentovskaya et al., 2011, Jones et al., 2010).

O-antigen polysaccharide chain is not present in *Y. pestis* which differentiates it from the LPS of other yersiniae. Although *Y. pestis* can also be able to produce common antigen of enterobacterial polysaccharide which is made up of trisaccharide repeating units which are comprised of a single residue of some of the compounds including *N*-acetyl- *D*-glucosamine (GlcNAc), 2-acetamido-2-deoxy-*D*mannuronic acid (ManNAcA), partially N-deacetylated and others (Knirel et al., 2012).

2.7.3. Lipid A from E. Coli, P. Putida, and P. Taiwanensis

Pseudomonas putida is required to produce various types of polyhydroxyalkanoate polymers and phospholipids. Membranous lipid A of P. putida plays an important role in stress condition (Wang et al., 2015). P. aeruginosa and P. putida belong to the same genus. Lipid A from E. coli, P. putida, and P. taiwanensis are gram-negative bacteria with LPS as a major part of the outer membrane. Reversed-phase HPLC was utilized for their analysis by associating with high-resolution MS. 7 molecules of lipid A from E.Coli were found through accurate mass measurements, 8 species *P. taiwanensis*, while 7 species in *P. putida* were observed. These lipid A species are primarily monophosphorylated and diphosphorylated with varied numbers and types of their substituted fatty acids. Acyl chains positions between *E. coli* and *Pseudomonas spp* also vary. Furthermore, the Hexa-acylated structure of *E.coli*-type consists of 4 fatty acids which are exchanged on the non-reducing end of the glucosamine backbone, whereas three acyl chains have been present on both sides *Pseudomonas*-type" Hexa-acylated species. Hepta, Hexa and Penta-acylated structures are the identified standard of acylation for *all 3*species (Froning et al., 2020). Lipid A of *E. coli* has a β(1-6)-linked backbone of glucosamine disaccharide along with distal glucosamine hydroxyl group which connects lipid A with the central region along with double phosphate groups. Furthermore, primary acyl chains are linked directly to the moieties of sugar whereas esterification of secondary acyl chains with the hydroxyl groups of primary acyl chains also done. All primary acyl chains of *E. coli* lipid are hydroxymyristates, among 2 secondary acyl chains one is myristate while the other one is laureate (Steimle et al., 2016).

2.7.4. S. Marcescens

Serratiamarcescens, a Gram-negative bacillus belongs to the Enterobacteriaceae family, found in the soil, water, and food with being unfavorable to the indigenous bacterium, although it is also called a human opportunistic pathogen due to its involvement in hospital-acquired infections (Makimura et al., 2007). *S. marcescens* is usually found as a red-pigmented bacterium due to prodigiosin production (Thomson et al., 2000), but bacteria which is reported in-hospital infections are non-pigmented (Carbonell et al., 2000). Lipid A characterization, showed that lipid A structure corresponds to the Diglu- cosamine ($GlcN_2$) backbone with diphosphates at 1 position on reducing side and the 4 position on the non-reducing side, a hydroxyl fatty acid at the 2 positions on the reducing side, and 2 acyloxyacyl groups at the 2 and 3 positions on the non-reducing side. S. marcescens lipid A is structurally different from compound 506, which is a classical lipid A of representative enterobacteria (Alexander & Za¨hringer, 2002) concerning the number of acyl chains. Acyl chain length and phosphate group present in lipid A, are considered as the prime factors required for endotoxic activity (Makimura et al., 007).

Genetically it was identified that the core region of the LPS from *S. marcescens* N28b has three common residues with *K. pneumonia.* Along with this, all known cores from LPS of *Enterobacteriaceae share* common inner core features. LPS Sugar analysis disclosed the availability of Glc, GlcN, GalA, Kdo, and other residues also (Makimura et al., 2007).

2.7.5. Desulfovibriodesulfuricans

Desulfovibriodesulfuricans bacterial species are sulfate-reducing Gram-negative rods that play a major role in the settlement of ecosystems lacking oxygen. However, *Desulfovibrio* species can grow in aerobic conditions. The polysaccharide chain of *D. desulfuricans* LPS consists of rhamnose, fucose, mannose, glucose, galactose, and heptose as the vital components in which galactose and rhamnose were identified as predominant carbohydrates in LPS structure, with 45.3% and 23.3% of all identified sugars. Whereas fucose is five times lower than rhamnose and mannose constitute very low about 4.2% only while glucose contribute 16.4% in carbohydrate profile. Moreover, the presence of Kdo and glucosamine in its structure is also identified through the derivatization of carbohydrates of investigated endotoxin to acetylated methyl glycosides. From GC/MS analysis, fatty acids in D. desulfuricans with the chain length of the C12–C18 were observed. 3-hydroxytetradecanoic acid (3-OH 14:0) was the primary fatty acid and its derivative produced methyl ester along with methyl esters of 3-metoxytetradecanoic acid and tetradecenoic acid. Furthermore, dodecanoic, tetradecanoic, and hexadecanoic acid methyl esters were present in considerable amounts among the compounds which are analyzed. Whereas the derivatives of other various fatty acid were detected in smaller amounts (Lodowska et al., 2012).

Conclusion

Lipid A is a major part of LPS or endotoxin that is involved in resistance against the antibiotic. The different methods used to isolate and characterize different bacteria show that lipid A structure changes with the type of bacterial strain and environment.

References

Alexander, C., & Zähringer, U. (2002). Chemical structure of lipid A-the primary immunomodulatory center of bacterial lipopolysaccharides. *Trends in Glycoscience and Glycotechnology*, *14*(76), 69-86.

Annamária, B. (2012). Structural characteristics of bacterial endotoxins.

Apicella, M. A. (2008). Isolation and characterization of lipopolysaccharides. *Bacterial Pathogenesis*, 3-13.

Bertani, B., & Ruiz, N. (2018). Function and biogenesis of lipopolysaccharides. *EcoSal Plus*, *8*(1).

Brodbelt, J. S. (2014). Photodissociation mass spectrometry: new tools for characterization of biological molecules. *Chemical Society Reviews*, *43*(8), 2757-2783.

Caboni, M., Pedron, T., Rossi, O., Goulding, D., Pickard, D., Citiulo, F., ...&Gerke, C. (2015). An O antigen capsule modulates bacterial pathogenesis in Shigellasonnei. *PLoS pathogens*, *11*(3), e1004749.

Carbonell, G. V., Della Colleta, H. H. M., Yano, T., Darini, A. L. C., Levy, C. E., & Fonseca, B. A. L. (2000). Clinical relevance and virulence factors of pigmented Serratiamarcescens. *FEMS Immunology & Medical Microbiology*, *28*(2), 143-149.

Davis Jr, M. R. (2013). *Lipopolysaccharide and Alginate in Mucoid Pseudomonas aeruginosa*. University of Virginia.

Davis Jr, M. R., & Goldberg, J. B. (2012). Purification and visualization of lipopolysaccharide from Gram-negative bacteria by hot aqueous-phenol extraction. *Journal of visualized experiments: JoVE*, (63).

Deisingh, A. K., & Thompson, M. (2004). Biosensors for the detection of bacteria. *Canadian journal of microbiology*, *50*(2), 69-77.

DeLaney, K., Sauer, C. S., Vu, N. Q., & Li, L. (2019). Recent advances and new perspectives in capillary electrophoresis-mass spectrometry for single cell "omics". *Molecules*, *24*(1), 42.

Dentovskaya, S. V., Anisimov, A. P., Kondakova, A. N., Lindner, B., Bystrova, O. V., Svetoch, T. E., ... & Knirel, Y. A. (2011). Functional characterization and biological significance of Yersinia pestis lipopolysaccharide biosynthesis genes. *Biochemistry (Moscow)*, *76*(7), 808-822.

Dentovskaya, S. V., Bakhteeva, I. V., Titareva, G. M., Shaikhutdinova, R. Z., Kondakova, A. N., Bystrova, O. V., ... & Anisimov, A. P. (2008). Structural diversity and endotoxic activity of the lipopolysaccharide of Yersinia pestis. *Biochemistry (Moscow)*, *73*(2), 192-199.

De Haas, C. J., van Leeuwen, H. J., Verhoef, J., van Kessel, K. P., & van Strijp, J. A. (2000). Analysis of lipopolysaccharide (LPS)-binding characteristics of serum components using gel filtration of FITC-labeled LPS. *Journal of immunological methods*, *242*(1-2), 79-89.

D'Haeze, W., Leoff, C., Freshour, G., Noel, K. D., & Carlson, R. W. (2007). Rhizobium etli CE3 bacteroid lipopolysaccharides are structurally similar but not identical to those produced by cultured CE3 bacteria. *Journal of Biological Chemistry*, *282*(23), 17101-17113.

El Hamidi, A., Tirsoaga, A., Novikov, A., Hussein, A., &Caroff, M. (2005). Microextraction of bacterial lipid A: easy and rapid method for mass spectrometric characterization. *Journal of lipid research*, *46*(8), 1773-1778.

Fàbrega, A., & Vila, J. (2012). Yersinia enterocolitica: pathogenesis, virulence and antimicrobial resistance. *Enfermedadesinfecciosas y microbiologiaclinica*, *30*(1), 24-32.

Farhana, A., & Khan, Y. S. (2021). Biochemistry, lipopolysaccharide. *StatPearls [Internet]*.

Fomsgaard, A., Freudenberg, M. A., &Galanos, C. (1990). Modification of the silver staining technique to detect lipopolysaccharide in polyacrylamide gels. *Journal of clinical microbiology*, *28*(12), 2627-2631.

Froning, M.; Helmer, P.O.; Hayen, H(2020). Identification and structural characterization of lipid A from Escherichia coli, Pesudomonas putida and Pseudomonas taiwanensisusing liquid chromatography coupled to high-resolution tandem mass spectrometry. *Rapid Commun. Mass Specrom.*, *34*, e8897.

Garcia, J., Bennett, D. H., Tancredi, D. J., Schenker, M. B., Mitchell, D. C., Reynolds, S. J., ...&Mitloehner, F. M. (2012). Characterization of endotoxin collected on California dairies using personal and area-based sampling methods. *Journal of occupational and environmental hygiene*, *9*(10), 580-591.

Girardin SE, Boneca IG, Carneiro LA, Antignac A, Jéhanno M, Viala J, Tedin K, Taha MK, Labigne A, Zähringer U, Coyle AJ, DiStefano PS, Bertin J, Sansonetti PJ, Philpott DJ. Nod1 detects a unique muropeptide from gram-negative bacterial peptidoglycan. Science. 2003 Jun 6; 300 (5625): 1584-7. PubMed PMID: 12791997. *infection*, *281*(17), 11637-48.

Gomes, J. M. G., de Assis Costa, J., & Alfenas, R. D. C. G. (2017). Metabolic endotoxemia and diabetes mellitus: a systematic review. *Metabolism*, *68*, 133-144.

Hardy, E., Pupo, E., Castellanos-Serra, L., Reyes, J., & Fernández-Patrón, C. (1997). Sensitive reverse staining of bacterial lipopolysaccharides on polyacrylamide gels by using zinc and imidazole salts. *Analytical biochemistry*, *244*(1), 28-32.

Hardy, E., Pupo, E., Santana, H., Guerra, M., & Castellanos-Serra, L. R. (1998). Elution of lipopolysaccharides from polyacrylamide gels. *Analytical biochemistry*, *259*(1), 162-165.

Henderson, J. C., O'Brien, J. P., Brodbelt, J. S., & Trent, M. S. (2013). Isolation and chemical characterization of lipid A from gram-negative bacteria. *JoVE (Journal of Visualized Experiments)*, (79), e50623.

Hirschfeld, M., Ma, Y., Weis, J. H., Vogel, S. N., & Weis, J. J. (2000). Cutting edge: repurification of lipopolysaccharide eliminates signaling through both human and murine toll-like receptor 2. *The Journal of Immunology*, *165*(2), 618-622.

Hübner, G., & Lindner, B. (2009). Separation of R-form lipopolysaccharide and lipid A by CE–Fourier-transform ion cyclotron resonance MS. *Electrophoresis*, *30*(10), 1808-1816.

Inzana, T. J., & Pichichero, M. E. (1984). Lipopolysaccharide subtypes of Haemophilus influenzae type b from an outbreak of invasive disease. *Journal of clinical microbiology*, *20*(2), 145-150.

Jauho, E. S., Boas, U., Wiuff, C., Wredstrøm, K., Pedersen, B., Andresen, L. O., ...& Jakobsen, M. H. (2000). New technology for regiospecific covalent coupling of polysaccharide antigens in ELISA for serological detection. *Journal of immunological methods*, *242*(1-2), 133-143.

Jeyaretnam, B. S. (1998). *Unusual rhizobial lipopolysaccharides: Their structures, biosynthesis, and biological activities*. University of Georgia.

Joly-Guillou, M. L. (2005). Clinical impact and pathogenicity of Acinetobacter. *Clinical microbiology and infection*, *11*(11), 868-873.

Jones, J. W., Cohen, I. E., Tureĉek, F., Goodlett, D. R., & Ernst, R. K. (2010). Comprehensive structure characterization of lipid A extracted from Yersinia pestis for determination of its phosphorylation configuration. *Journal of the American Society for Mass Spectrometry*, *21*(5), 785-799.

Kalambhe, D. G., Zade, N. N., & Chaudhari, S. P. (2017). Evaluation of two different lipopolysaccharide extraction methods for purity and functionality of LPS. *Int J CurrMicrobiolApplSci*, *6*(3), 1296-1302.

Kido, N. O. B. U. O., Ohta, M. I. C. H. I. O., & Kato, N. O. B. U. O. (1990). Detection of lipopolysaccharides by ethidium bromide staining after sodium dodecyl sulfate-polyacrylamide gel electrophoresis. *Journal of bacteriology*, *172*(2), 1145-1147.

Knirel, Y. A., &Anisimov, A. P. (2012). Lipopolysaccharide of Yersinia pestis, the cause of plague: structure, genetics, biological properties. *ActaNaturae (англоязычнаяверсия)*, *4*(3 (14)).

Knirel, Y. A., Kondakova, A. N., Bystrova, O. V., Lindner, B., Shaikhutdinova, R. Z., Dentovskaya, S. V., & Anisimov, A. P. (2008). New features of Yersinia lipopolysaccharide structures as revealed by high-resolution electrospray ionization mass spectrometry. *Advanced Science Letters*, *1*(2), 192-198.

Knirel, Y. A., Lindner, B., Vinogradov, E. V., Kocharova, N. A., Senchenkova, S. Y. N., Shaikhutdinova, R. Z., ... & Anisimov, A. P. (2005). Temperature-dependent variations and intraspecies diversity of the structure of the lipopolysaccharide of Yersinia pestis. *Biochemistry*, *44*(5), 1731-1743.

Komuro, T., & Galanos, C. (1988). Analysis of Salmonella lipopolysaccharides by sodium deoxycholate—polyacrylamide gel electrophoresis. *Journal of Chromatography A*, *450*(3), 381-387.

Kumar, H., Kawai, T., & Akira, S. (2011). Pathogen recognition by the innate immune system. *International reviews of immunology*, *30*(1), 16-34.

Kutschera, A., Schombel, U., Schwudke, D., Ranf, S., & Gisch, N. (2021). Analysis of the Structure and Biosynthesis of the Lipopolysaccharide Core Oligosaccharide of Pseudomonas syringaepv. tomato DC3000. *International journal of molecular sciences*, *22*(6), 3250.

Leimbach, A., Hacker, J., & Dobrindt, U. (2013). E. coli as an all-rounder: the thin line between commensalism and pathogenicity. *Between pathogenicity and commensalism*, 3-32.

Lesse, A. J., Campagnari, A. A., Bittner, W. E., & Apicella, M. A. (1990). Increased resolution of lipopolysaccharides and lipooligosaccharides utilizing tricine-sodium dodecyl sulfate-polyacrylamide gel electrophoresis. *Journal of immunological methods*, *126*(1), 109-117.

Li, J., Cox, A. D., Hood, D. W., Schweda, E. K., Moxon, E. R., & Richards, J. C. (2005). Electrophoretic and mass spectrometric strategies for profiling bacterial lipopolysaccharides. *Molecular BioSystems*, *1*(1), 46-52.

Li, J., Cox, A. D., Hood, D., Moxon, E. R., & Richards, J. C. (2004). Application of capillary electrophoresis-electrospray-mass spectrometry to the separation and characterization of isomeric lipopolysaccharides of Neisseria meningitidis. *Electrophoresis*, *25*(13), 2017-2025.

Li, J., Dzieciatkowska, M., Hood, D. W., Cox, A. D., Schweda, E. K., Moxon, E. R., & Richards, J. C. (2007). Structural characterization of sialylatedglycoforms of H. influenzae by electrospray mass spectrometry: fragmentation of protonated and sodiated O-deacylated lipopolysaccharides. *Rapid Communications in Mass*

Spectrometry: An International Journal Devoted to the Rapid Dissemination of Up-to-the-Minute Research in Mass Spectrometry, *21*(6), 952-960.

Livorsi, D. J., MacNeil, J. R., Cohn, A. C., Bareta, J., Zansky, S., Petit, S., ...& Farley, M. M. (2012). Invasive Haemophilus influenzae in the United States, 1999–2008: epidemiology and outcomes. *Journal of Infection*, *65*(6), 496-504.

Lodowska, J., Wolny, D., Jaworska-Kik, M., Kurkiewicz, S., Dzierżewicz, Z., &Węglarz, L. (2012). The chemical composition of endotoxin isolated from intestinal strain of Desulfovibriodesulfuricans. *The Scientific World Journal*, *2012*.

Lukasiewicz, J., Niedziela, T., Jachymek, W., Kenne, L., &Lugowski, C. (2006). Structure of the lipid A–inner core region and biological activity of Plesiomonasshigelloides O54 (strain CNCTC 113/92) lipopolysaccharide. *Glycobiology*, *16*(6), 538-550.

Madsen, J. A., Cullen, T. W., Trent, M. S., &Brodbelt, J. S. (2011). IR and UV photodissociation as analytical tools for characterizing lipid A structures. *Analytical chemistry*, *83*(13), 5107-5113.

Magdaleno, F., Blajszczak, C. C., & Nieto, N. (2017). Key events participating in the pathogenesis of alcoholic liver disease. *Biomolecules*, *7*(1), 9.

Mahajan-Miklos, S., Tan, M. W., Rahme, L. G., &Ausubel, F. M. (1999). Molecular mechanisms of bacterial virulence elucidated using a Pseudomonas aeruginosa–Caenorhabditis elegans pathogenesis model. *Cell*, *96*(1), 47-56.

Makimura, Y., Asai, Y., Sugiyama, A., & Ogawa, T. (2007). Chemical structure and immunobiological activity of lipid A from Serratiamarcescens LPS. *Journal of medical microbiology*, *56*(11), 1440-1446.

Mather, A. E., Reid, S. W. J., Maskell, D. J., Parkhill, J., Fookes, M. C., Harris, S. R., ... & Thomson, N. R. (2013). Distinguishable epidemics of multidrug-resistant Salmonella Typhimurium DT104 in different hosts. *Science*, *341*(6153), 1514-1517.

Munford, R. S. (2008). Sensing gram-negative bacterial lipopolysaccharides: a human disease determinant?. *Infection and immunity*, *76*(2), 454-465.

Phillips, N. J., Schilling, B., McLendon, M. K., Apicella, M. A., & Gibson, B. W. (2004). Novel modification of lipid A of Francisellatularensis. *Infection and immunity*, *72*(9), 5340-5348.

Prior, J. L., Prior, R. G., Hitchen, P. G., Diaper, H., Griffin, K. F., Morris, H. R., ... & Titball, R. W. (2003). Characterization of the O antigen gene cluster and structural analysis of the O antigen of Francisellatularensis subsp. tularensis. *Journal of medical microbiology*, *52*(10), 845-851.

Pupo, E., López, C. M., Alonso, M., & Hardy, E. (2000). High-efficiency passive elution of bacterial lipopolysaccharides from polyacrylamide gels. *ELECTROPHORESIS: An International Journal*, *21*(3), 526-530.

Que, N. L., Lin, S., Cotter, R. J., & Raetz, C. R. (2000). Purification and mass spectrometry of six lipid A species from the bacterial endosymbiont Rhizobium etli: demonstration of a conserved distal unit and a variable proximal portion. *Journal of Biological Chemistry*, *275*(36), 28006-28016.

Qureshi, N., Kaltashov, I., Walker, K., Doroshenko, V., Cotter, R. J., Takayama, K., ...&Golenbock, D. T. (1997). Structure of the monophosphoryl lipid A moiety obtained from the lipopolysaccharide of Chlamydia trachomatis. *Journal of Biological Chemistry*, *272*(16), 10594-10600.

Rebeil, R., Ernst, R. K., Gowen, B. B., Miller, S. I., &Hinnebusch, B. J. (2004). Variation in lipid A structure in the pathogenic yersiniae. *Molecular microbiology*, *52*(5), 1363-1373.

Rebeil, R., Ernst, R. K., Jarrett, C. O., Adams, K. N., Miller, S. I., &Hinnebusch, B. J. (2006). Characterization of late acyltransferase genes of Yersinia pestis and their role in temperature-dependent lipid A variation. *Journal of bacteriology*, *188*(4), 1381-1388.

Sarmikasoglou, E., &Faciola, A. P. (2021). Ruminal Lipopolysaccharides Analysis: Uncharted Waters with Promising Signs. *Animals*, *11*(1), 195.

Schilling, B., McLendon, M. K., Phillips, N. J., Apicella, M. A., & Gibson, B. W. (2007). Characterization of lipid A acylation patterns in Francisellatularensis, Francisellanovicida, and Francisellaphilomiragia using multiple-stage mass spectrometry and matrix-assisted laser desorption/ionization on an intermediate vacuum source linear ion trap. *Analytical chemistry*, *79*(3), 1034-1042.

Stead, C., Tran, A., Ferguson Jr, D., McGrath, S., Cotter, R., & Trent, S. (2005). A novel 3-deoxy-D-manno-octulosonic acid (Kdo) hydrolase that removes the outer Kdo sugar of Helicobacter pylori lipopolysaccharide. *Journal of bacteriology*, *187*(10), 3374-3383.

Steimle, A., Autenrieth, I. B., & Frick, J. S. (2016). Structure and function: Lipid A modifications in commensals and pathogens. *International Journal of Medical Microbiology*, *306*(5), 290-301.

Stephens, M., Liao, S., & von der Weid, P. Y. (2021). Ultra-purification of Lipopolysaccharides reveals species-specific signalling bias of TLR4: importance in macrophage function. *Scientific reports*, *11*(1), 1-11.

Stromberg, L. R., Mendez, H. M., &Mukundan, H. (2017). Detection methods for lipopolysaccharides: past and present. *Escherichia coli-recent advances on physiology, pathogenesis and biotechnological applications. InTech*, 141-168.

Su, W., & Ding, X. (2015). Methods of endotoxin detection. *Journal of laboratory automation*, *20*(4), 354-364.

Subhi, I. M., Zgair, A. K., &Ghafil, J. A. (2017). Extraction and Purification of Pseudomonas aeruginosa.

Thomas, R. M., Titball, R. W., Oyston, P. C., Griffin, K., Waters, E., Hitchen, P. G., ... & Prior, J. L. (2007). The immunologically distinct O antigens from Francisellatularensis subspecies tularensis and Francisellanovicida are both virulence determinants and protective antigens. *Infection and immunity*, *75*(1), 371-378.

Thomson, N. R., Crow, M. A., McGowan, S. J., Cox, A., & Salmond, G. P. C. (2000). Biosynthesis of carbapenem antibiotic and prodigiosin pigment in Serratia is under quorum sensing control. *Molecular microbiology*, *36*(3), 539-556.

Vilches, S., Canals, R., Wilhelms, M., Salo, M. T., Knirel, Y. A., Vinogradov, E., ...& Tomas, J. M. (2007). Mesophilic Aeromonas UDP-glucose pyrophosphorylase (GalU) mutants show two types of lipopolysaccharide structures and reduced virulence. *Microbiology*, *153*(8), 2393-2404.

Vinogradov, E., & Perry, M. B. (2004). Characterisation of the core part of the lipopolysaccharide O-antigen of Francisellanovicida (U112). *Carbohydrate research*, *339*(9), 1643-1648.

Vinogradov, E., Perry, M. B., &Conlan, J. W. (2002). Structural analysis of Francisellatularensis lipopolysaccharide. *European journal of biochemistry*, *269*(24), 6112-6118.

Wang, X., & Quinn, P. J. (Eds.). (2010). *Endotoxins: structure, function and recognition* (Vol. 53). Springer Science & Business Media.

Wang, X., Ribeiro, A. A., Guan, Z., McGrath, S. C., Cotter, R. J., &Raetz, C. R. (2006). Structure and biosynthesis of free lipid A molecules that replace lipopolysaccharide in Francisellatularensis subsp. novicida. *Biochemistry*, *45*(48), 14427-14440.

Wang, X., Zhang, C., Shi, F., & Hu, X. (2010). Purification and characterization of lipopolysaccharides. In *Endotoxins: Structure, Function and Recognition* (pp. 27-51). Springer, Dordrecht.

Wang, Y., Wang, J., Li, Y., Wang, B., Tao, G., & Wang, X. (2015). Structure characterization of phospholipids and lipid A of Pseudomonas putida KT2442. *European journal of mass spectrometry*, *21*(5), 739-746.

Yamaoka, Y., & Graham, D. Y. (2014). Helicobacter pylori virulence and cancer pathogenesis. *Future oncology*, *10*(8), 1487-1500.

Yokota, S. I., Amano, K. I., Shibata, Y., Nakajima, M., Suzuki, M., Hayashi, S., ...&Yokochi, T. (2000). Two distinct antigenic types of the polysaccharide chains of Helicobacter pylori lipopolysaccharides characterized by reactivity with sera from humans with natural infection. *Infection and immunity*, *68*(1), 151-159.

Zhou, Z., Lin, S., Cotter, R. J., &Raetz, C. R. (1999). Lipid A Modifications Characteristic of Salmonella typhimurium Are Induced by NH4VO3 inEscherichia coli K12*: DETECTION OF 4-AMINO-4-DEOXY-l-ARABINOSE, PHOSPHOETHANOLAMINE AND PALMITATE. *Journal of Biological Chemistry*, *274*(26), 18503-18514.

Chapter 3

Endotoxin Biosynthesis: Genetics and Biochemistry of the Process

Prakriti Sharma[1], Chetna Mahajan[2], Abhishek Gupta[3] and M. P. S. Tomar[4,*]

[1]College of Animal Biotechnology, Guru Angad Dev Veterinary and Animal Sciences University, Ludhiana, Punjab, India
[2]College of Veterinary Science, Rampuraphul, GADVASU, India
[3]Post Graduate Institute of Veterinary Education and Research (PGIVER), Jaipur, India
[4]N.T.R. College of Veterinary Science, Gannavaram, India

Abstract

Endotoxin at a molecular level called LPS presents in bacterial outer cell envelope composed of lipid A, a short core oligosaccharide, and the O-antigen polysaccharide. Lipid A is required for bacterial growth to maintain the integrity of the outer membrane barrier and is the principal trigger of the immune response. Variation in the structure of Lipid A and biosynthesis alters bacteria's resistance and growth system. Hence, it is essential to understand the biosynthesis of endotoxin and the mechanism of virulence and antibiotic resistance to find a new target for antibacterial treatments. Genes such as lpxA, lpxB, lpxC and lpxD are involved in regulating Lipid A biosynthesis pathway. However, the available data suggest that bacteria have variations in the Lipid A biosynthetic pathways that resist the bacteria against antibiotics and enhance bacterial survival. This chapter comprises the mechanism of endotoxin biosynthesis and genetic variations that give different characterisation and properties to the bacterial endotoxins.

Keywords: endotoxin, LPS, lipid A

---

* Corresponding Author's E-mail: anatomistpdtr@gmail.com.

In: Endotoxins and their Importance
Editors: Arif Pandit and R. S. Sethi
ISBN: 978-1-68507-839-3

Introduction

Lipopolysaccharides (LPS) are present on the cell membrane of gram-negative bacteria that elicit an immune response in humans following bacterial infection. The bacterial cell wall comprises the outer membrane, peptidoglycan layer, and inner phospholipid bilayer. Endotoxins are present in the outer membrane. They are released in the air after the destruction of the bacterial cell wall (Sampath, 2018). Endotoxins are composed of Lipid A, oligosaccharide, and an o- specific chain (O- specific polysaccharides). But Corsaro et al., (2008) observed that LPS is present in two forms a) smooth form with three regions viz. Lipid A/ endotoxin means Lipid A is the endotoxin, oligosaccharide core region and o- specific polysaccharides, b) Rough- form is also known as lipooligosaccharides (LOSs) that do not have a polysaccharide chain. The peripheral surface of the outer membrane has glucosamine derived phospholipid 4 known as Lipid A, responsible for the toxic activity of endotoxin and has antigenic properties (Rietschel et al., 1994). O-chain determined the surface antigen form of oligosaccharides and is different for different strains (Gorbet and Sefton, 2005). Lipid A and polysaccharide side chains of LPS are responsible for virulent behaviour in gram-negative bacteria (Sampath, 2018). These three components of LPS affect endotoxin's structure and morphology (Whitfield et al., 1997). *E. coli* have 106 Lipid and 107 glycerophospholipids. Lipid A also plays a role in the growth of *E. coli* cells with the wild type cells having core and o antigens (Raetz& Whitfield, 2002. The inner core of LPS is conserved while the Lipid A region has the variable portion. All the gram-negative strains do not have complete LPS but have the minimum required Lipid A with one kda residue necessary for the growth of bacterial cells (Helander et al., 1988). Gram-negative bacteria show multidrug resistance. Therefore, new pathways should be studied to avoid this problem of antibiotic resistance. Endotoxin with toxic properties is responsible for bacteria mediated diseases.

Delta endotoxin (Δ endotoxin) is another type of endotoxin beneficial and used as insecticide. It is encoded by cry genes present in the plasmid of *Bacillus thuringiensis*. During sporulation cry genes are transcribed by RNA polymerase and produce protoxins. These are effective against insects and are bioinsecticides. Δendotoxin were synthesised during sporulation as prototoxins enveloped by intracellular inclusion bodies. When an insect ingests the prototoxin, it leads to its transformation into toxin by gut enzymes such as trypsin and insect death (Aronson, 2002).

Lipid A an active component of LPS and biosynthesis of Lipid A can be targeted to get rid of resistance-related problems. LpxC enzyme catalyses Lipid A biosynthesis's committed step can be targeted. LpxC-4, an inhibitor for LpxC, can be used. But a spontaneous mutation in the bacteria for this inhibitor frequently occurs compared to other antibiotics. But still, lpxC-4 is effective for various strains of gram-negative bacteria and thus can be used to target the pathogen. Because of the resistance of gram-negative bacteria, the biosynthesis and regulatory mechanism of endotoxin should be studied (Tomaras et al., 2014)

3.1. Lipid A

3.1.1. Biosynthesis

Lipid A is the active component of endotoxin with hydrophobic nature. Its synthesis occurs at the interface between the cytosol and the inner membrane. It has 6 to 7 saturated fatty acids attached to the dimer of phosphorylated N-acetyl glucosamine (Sampath, 2018). Lipid A is glucosamine β (1-6) 1- 4'-biphosphate disaccharide having acyl group at 2, 3 and 2' 2' with 3' hydroxyl 3' hydroxyl myristate. Further fatty acyl chains have ester group given by laurate and myristate at 2 and 3' 3' hydroxyl group and 6' 6' position of Lipid A is glycosylated with two 3-deoxy- D-manno-octulosonic acid (KDO) (Trent et al., 2006). E. coli contain approximately 106 Lipid A and 107 glycerophospholipids.

Lipid A is the toxic part of LPS having phosphoryl group attached at '1'and '4'β (1-6)- linked D- glucosamine which is the general constituent of Lipid A. Most of the β (1-6) D- glucosamine have (R)-3-acyloxy and (R)-3-hydroxyacyl attached esters and amide groups. In Lipid A of enterobacteria, the 3- deoxy-D-mannooctulosonic and (R)-3-(tetradecanoyloxy) tetradeconic acid are attached to the D- glucosamine disaccharides at 3 and '3'position in place of hydroxyl groups (Rietschel et al., 1984). While in the Salmonella phosphate group and 2-keto-3deoxyoctonate are attached to β (1-6) D glucosamine at 4th and 3rd position (Gmeiner et al. 1971).The Lipid A structure of *Campylobacter jejuni* has two chains; one is hydrophilic, and the other is hydrophobic. The hydrophilic chain is of sugar composed of three (1 → 6)-associated bisphosphorylated hexosamine polar head groups disaccharides with 1:6:1.2ratio of D-glucosamine disaccharide: a combination of 2,3-diamino-2,3-dideoxy-D-glucose, and D-glucosamine and D-glucosamine: 3-

diamino-2,3-dideoxy-D-glucose. Phosphorylethano-lamine bound to the reducing sugar of hydrophilic chain and the non-reducing sugar at 4th position a diphosphoryl ethanolamine or an ester bound phosphate is bind. Study shows that '6'position is where polysaccharides are attached to the LPS. The hydrophobic part has fatty acids associated with six ester and amides molecules. Lipid A backbone of non-reducing sugar attached to four (*R*)-3-hydroxytetradecanoic acid at 2, 3, '2'and '3'position have two molecules of hexadecenoic acid or compound of one tetra decanoic and one of hexadecenoic acid at 3- hydroxyl group. The above study shows that *Campylobacter jejuni* have three backbones present in the Lipid A differ in amino sugar arrangement with the attachment of hybrid sugar and is different from other bacteria (MORAN et al., 1991).

Lipid A biosynthetic pathway is also referred to as "The Raetz Pathway". Bioinformatic studies said that although the Lipid A pathway is conserved, some gram-negative bacteria alter their Lipid A structure. Anderson et al., (1985) found that free extract obtained from gram negative bacteria E. coli has all enzymes required for acylation of UDP-GlcNAc and conversion to disaccharide 1-phosphate without energy sources (ATP/ phosphor enol pyruvate) from outside.

The first step of Lipid A synthesis is catalysed by the important enzyme also required for cell growth UDP-*N*-acetylglucosamine 3-*O*-acyltransferase encoded by gene LpxA followed by the lpxC gene-encoded enzyme deacetylase (Wyckoff et al., 1998). LpxA is a homotrimer having three active sites per trimer. LpxA induces the first reaction involving the UDP-GlcNAc acylation at sugar nucleotide (Raetz, 1993) and formation of UDP-3-O-(R-3-hydroxymyristoy1)-GlcNAc where R-3 hydroxymyristoyl attached to UDP-GlcNAc at '3'OH and ACP (acyl carrier protein) from UDP-GlcNAc with R-3-hydroxymyristoyl-ACP in the presence of enzyme LpxA. In the different bacteria, LpxA incorporates hydroxy acyl chain of different lengths. For example, in E. coliLpxA include C14 hydroxy acyl chain, Pseudomonas aeruginosa C10 hydroxyacyl chains, and Neisseria meningitides Leptospira interrogans C12 hydroxyacyl chains (Raetz et al., 2007). But this first step of Lipid A synthesis is reversible and unfavourable because of the equilibrium constant, approximately 0.01 (Anderson et al., 1993). Therefore Zinc-dependent metalloamidase enzyme UDP-3-O-(R-3-hydroxymyristoyl)-N-acetylglucosamine deacetylase or LpxC is an important enzyme that plays a significant role in Lipid A biosynthesis

LpxC catalyses it encoded UDP-3-O-(R-3-hydroxymyristoyl)-N-acetylglucosamine deacetylase enzyme that involved in deacetylation of UDP-

3-O-(R-3-hydroxymyristoy1)-GlcNAc (myr-UDP-GlcNAc) and form UDP3-O-(R-3-hydroxymyristoy1)-GlcN (myr-UDP-GlcN). This step is a favourable and committed step of Lipid A synthesis. A single copy of the enzyme coding gene is conserved in all grams –ve bacteria (Zhou & Barb, 2008 and Clayton et al., 2013). LpxC of *Aquifexaeolicus* has α+ β fold with Zinc binding motifs. Here Zn ion is present at the base of the active site of an enzyme and near to hydrophobic tunnel. This tunnel is responsible for enzyme specificity. Therefore, an inhibitor against this tunnel can be used as an antibacterial drug (Whittington et al., 2003).

The third step involves the acylation of UDP3-O-(acyl)-GlcN. LpxD encoded *N*-acyltransferase is the enzyme that adds a second acyl chain to the free amino group of deacylated UDP3-O-(acyl)-GlcN and forms UDP-2,3-diacylGlcN (Whitfield & Trent, 2014 and Bartling & Raetz, 2008). LpxD is present as a homotrimer, meaning three identical subunits. Each subunit fuses uridine binding domain at the N terminal, helical extension at the C terminal, and a lipid-binding core. In this enzyme Phe-43 and Tyr-49 bind with uracil by quadruple interaction while Asn-46 and His-284 attached to phosphate group by hydrogen bond and set the glucosamine of Lipid A in the centre. Fatty acid-binding side is present near the catalytic centre His-247 and His-284 help in the acylation of UDP3-O-(acyl)-GlcN where they are involved in the amine mediated nucleophilic attack on the thioester conjugated acyl carrier protein (Buetow et al., 2007). LpxD also play a role in the growth of biofilm. Mutation in the lpxD results in a decrease in biofilm formation and bacterial adherence to the epithelial cells of the airways. So, this can be a target for treatment as alteration of lpxD expression in *P. aeruginosa* affects the ability of bacterial infection by inhibiting the formation of biofilm (Alshalchi & Anderson, 2015).

In E. coli UDP-2,3-diacylglucosamine hydrolase encoded by ybbf gene denoted as LpxH involved in the hydrolysis of pyrophosphate bond of the product form by LpxD, i. e. UDP-2,3-diacylGlcN.It attacks UDP-2,3-diacylGlcN at αP atom with water and forms 2,3-diacylglucosamine 1-phosphate with 1 UMP (Babinski et al., 2002). Mutation of LpxH results in the accumulation of UDP-2,3-diacylGlcN. Approximately 50%-gram negative bacteria do not have clear LpxH orthologous previously in E. coli it was denoted as ybbF. In contrast, some bacteria lack LpxH contain weak orthologs of LpxH known as LpxH2. LpxH is present in E. coli, H. Influenza, Neisseria meningitides, Pasteurella multocida, Salmonella enteric, Salmonella typhimurium, Xylella fastidiosa and Yersinia pestis and bacteria having LpxH2 are Agrobacterium tumefaciens, Caulobacter crescentus, and

Sinorhizobiummeliloti. Some gram-negative bacteria such as Pseudomonas aeruginosa and Ralstonia solanacearum have orthologs of LpxH and LpxH2, and some such as *Aquifexaeolicus, Chlamydophila pneumonia, Chlamydia muridarum, and Rickettsia conorii*do not have LpxH and H2 ortholog (Babinski et al., 2002). Gram-negative bacteria lacking LpxH have different UDP-2,3-diacylglucosamine pyrophosphatase known as LpxI that form 2,3-diacylglucosamine 1-phosphate but by a different pathway. LpxI was found in *Caulobacter crescentus* known as cclpxI located between lpxA and lpxB. It catalyses the attack of water on β-Phosphate of UDP-2,3-diacylGlcN instead of α Phosphates catalyse by LpxH (Metzger et al., 2010).

Next step in synthesising Lipid A is catalysed by Lipid A disaccharide synthase that forms (β, '1'- 6) tetraacyl-disaccharide-1- phosphate (DSMP) from the reaction of UDP-2,3- diacyl glucosamine with 2,3-diacylglucosamine-1-phosphate. Lipid A disaccharide synthase is a glycosyltransferase that forms 2',3'-diacylglucosamine (β, '1'- 6), 2,3-diacylglucosamine-1- phosphate along with one UDP. In E. coli this Lipid A disaccharide synthase is encoded by the LpxB gene present immediately to LpxB and at 631 base pair of dnaE in a counterclockwise direction, and mRNA expression of LpxB and LpxA occurs in the clockwise direction (Crowell et al., 1986). Metzger and Raetz found that Lpx B is a conserved enzyme of the Lipid A synthesis pathway present in the peripheral surface of the membrane and required for E. coli. It was first found in the mutant strain of E. coli (pgsB1), which was temperature-sensitive (Bohl et al., 2018).

Lipid A'4'kinase encoded by the gene orfE, which is now denoted as LpxK perform the sixth step. It phosphorylates tetra acyl disaccharide 1-phosphate at the fourth position and forms tetra acyl disaccharide 1,4-bis-phosphate. This tetra acyl disaccharide 1,4'-bis-phosphate is also termed as Lipid 4_A. This Lipid A '4' kinase is also required for the growth of cells. Cells with a mutation in this gene do not survive (Garrett et al., 1997). LpxK faces the cytosolic side of the inner membrane (Emptage et al., 2013).

The next step involves transferring 3-deoxy-D-manno-octulosonic acid (KDO) an anionic sugar to Lipid $4_{A.}$ But before the incorporation of KDO to Lipid $4_{A,}$ CMP- KDO synthase activate KDO to CMP- KDO. In Escherichia coli and *Salmonella typhimurium,* inhibition of activation or synthesis of activated CMP- KDO results in the accumulation of Lipid 4_A and blocks the cell growth, shows its significant role in cell growth. This transfer of CMP-KDO to Lipid 4_A is required for the assembly of endotoxin. (Clementz & Raetz, 1991). The synthesis and incorporation of CMP- KDO to the Lipid 4_A were done with the help of enzymes such as KDO-8-phosphate synthetase,

KDO-8-phosphate phosphatase, CMP-KDO synthetase, and KDO-Lipid A transferase. Mutation in the KdsA gene that codes for KDO-8-phosphate synthetase inhibit bacterial growth due to the blockage of LipidA biosynthesis. *Salmonella typhimurium* can use this kdsA exogenously if it has this enzyme mutation, leading to cell growth (Goldman & Devine, 1987). CMP- KDO synthetase is encoded by kdsB gene activate KDO, an8-carbon sugar compound and form CMP- KDO from CTP and KDO in the presence of Mg^{+2} as a cofactor, which is then incorporated to the Lipid 4_A (Goldman et al., 1986). KDO- Lipid A transferase encoded by WaaA is required to transfer KDO sugar to Lipid $4_{A,}$ a Lipid A precursor. Different bacteria have KDO-Lipid A transferase with different properties such as bifunctional, mono-functional and trifunctional in which bifunctional transferase transfer two KDO sugars while mono-functional transfer one KDO sugar to one Lipid 4_A and three KDO sugar by trifunctional transferase. E. coli have KDO- Lipid A transferase encoded by gene kdtA with bifunctional property and Haemophilus influenza with mono-functional and Chlamydia trachomatis with trifunctional property (Hankins & Trent, 2009). In the H. influenza, KdtA encodes for mono-functional kdo transferase that transfers the single KDO to Lipid IV A. kdtA shows 50% similarity with E. coli one. kdo transferase reaction depends on CMP-kdo system (White et al., 1997).

There are three acyltransferases orthologs found in V. Cholera (Hankins & Trent, 2009). V. cholera requires a phosphate group for the secondary acylation of KDO residue. In contrast, Bordetella pertussis, Escherichia coli, and Haemophilus influenza require secondary substituent (additional Kdo sugar or a phosphate group) for KDO acylation catalysed by LpxL.

This step involves the addition of laurate and myristate to the KDO- Lipid 4_A. Enzymes involved in the addition of laurate and myristate are lauryl transferase present in the cytosol and myristoyl transferase present as membrane-associated. Gene encodes for lauryl transferase is htrB and for myristoyl transferase is msbB. Lauryl transferase has specific functions that add specificity to their work, such as it recognises only KDO linked Lipid IV, not the alone lipid IV and is typical for laurate. E. coli and S. typhimurium acyl chains have four R3-hydroxymyristates and one laurate and myristate (Brozek & Raetz, 1990). The donor for the addition of laurate and myristate are laurate Acyl carrier protein (ACP1) or myristate (ACP) (Clementz et al., 1996). In E. coli, two additional enzymes with the transferases may present that acylate (Kdo)2-lipid IVA (Clementz et al., 1997).

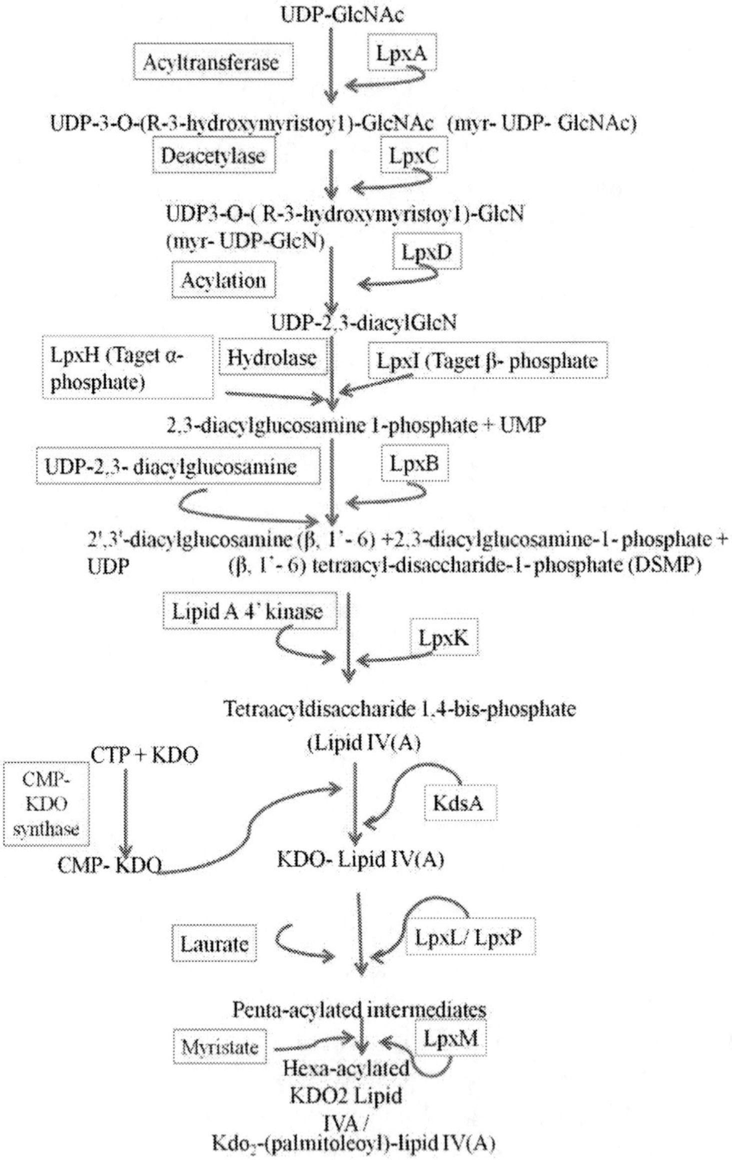

Figure 3.1. Lipid A biosynthesis pathway

KDO is also required to function secondary acyltransferases such as Lpxl, LpxM and LpxP. LpxP in the E. coli uses acyl- acyl carrier proteins (acyl-ACPs) as a donor, activated at low temperature (<200C), replaces laurate (C12:0) from 2nd position of the KDO2 Lipid IVA with palmitoleate (C16:1).

Where LpxL transferase first transfers laurate (C12:0) to the hydroxyl group of 2′-N-linked (R)-3-hydroxymyristate but at low-temperature LpxP transfer palmitoleate (C16:1) at the same position and form penta-acylated intermediate. Then LpxM transfers myristate (C16:0) to the penta- acylated Lipid A at free hydroxyl (OH) of 3′-O-linked (R)-3-hydroxymyristate. It was also studied that LpxL can also use acyl-coenzyme A (Brozek & Raetz, 1990 and Hankins & Trent, 2009).

In V. Cholera, three late acyltransferases have been found named Vc0212, Vc0213, and Vc1577 where only Vc0213 acts as myristoyl transferase and uses myristoyl-CoA as a donor of acyl (Hankins & Trent, 2009).

Pseudomonas aeruginosa has an enzyme lauryl transferase that transfer lauryl-acyl carrier protein (ACP) to lipid IV A does not need KDO attachment. The study shows that all gram-negative bacteria do not require KDO incorporation to lipid IV A for the attachment of lauryl (Mohan & Raetz, 1994).

3.1.2. Lipid A Remodelling

The process of Lipid A synthesis is conserved in nature, but different organism shows diversity in the Lipid A. Bacteria uses this strategy to promote their growth, resistance against the innate immune system and antibiotic and increase the virulence (Gattis, et al., 2013). Single *C. trachomatis* gene (gseA) isolated from *Chlamydia trachomatis* can form a gene-specific epitope and induce this character in E. coli when cloned. It is 23% identical to kdtA (an enzyme that adds KDO to Lipid A). E. coli with gseA gene have three KDO attached, two added by kdtA and one by gseA. gseA gene is trifunctional and adds three KDO to Lipid A in the E. coli having mutation for kdtA gene because KDO is very important for LPS synthesis. Therefore, inhibiting LPS synthesis by targeting gseA gene can be used for treatment against *Chlamydia trachomatis* (Belunis et al., 1992).

Lipid A modification is essential for bacterial survival in the stress; therefore, enzymes involved in Lipid A modification are regulated at transcriptional and post-translational levels by small RNAs, peptides, and substrate availability. Alteration of the bacterial membrane by remodelling of Lipid A alters the pathogenesis. Modifications in the Lipid A structure involves in the virulence of bacteria. It is a strategy of gram-negative bacteria by which it receives antibiotic resistance that hampers the host immune system's ability to recognise the LPS/ endotoxin (Needham & Trent, 2013).

The outer membrane of Francisella by alteration of Lpx D enzyme may be a process of bacterial survival. In Francisella, two types of LpxD genes LpxD1 and LpxD2 are present. Temperature-dependent study shows that LpxD1 adds longer acyl chain with the 3-OH at acyl group of C18 at 37oresult in bacterial survival in the mammalian host while LpxD2 18oC at adds the same 3-OH at C16 with the shorter acyl chain to the Lipid A and help bacteria to survive in the cold-blooded host (Li et al., 2012).

Mutation in *Salmonella typhimurium* produces imperfect 2-keto-3-deoxyoctonate that affects growth and the LPS synthesis because of this mutation in S. Typhimurium KDO-8-'P'synthetase become temperature sensitive. Also, under some conditions where KDO synthesis is affected, the imperfect Lipid A precursor becomes accumulated. This flawed Lipid A comprises glucosamine disaccharides with phosphate and three hydroxy myristate. Out of three hydroxy myristate one has the ester linkage while the other has amide linkage. This new compound also plays a role as an intermediate or Lipid A precursor for synthesising LPS (Rick et al., 1977).

PhoQ is a sensor kinase that spans the inner membrane of bacteria. The cytoplasmic domain of PhoQ has the histidine kinase that trans-autophosphorylates in response to the environment and further phosphorylates PhoP at aspartate and activate it. PhoP bind the responsive DNA region and regulates the expression. The sensor kinase in *Salmonella typhimurium's* regulation system is the PhoPQ system that regulates the expression of many genes required for virulence, growth, antimicrobial resistance, and Lipid A synthesis. Activation of PhoQ occurs in the phagosome and host tissue in response to antibiotics, CAMP, and acidic pH. Thus, this PhoPQ system is involved in the pathogenesis of S. *typhimurium* and other bacteria (Prost & Miller, 2008). Miller et al., 1989 studied that PhoP mutant strain of *Salmonella typhimurium* shows depletion of virulence having PhoP mutation. The strains with a mutation in pagC, PhoQ and PhoP do not survive. Therefore, a mutation in PhoP and PhoQ can be used for the development of an attenuated live vaccine.

PhoP- PhoQ system also provide host cationic antimicrobial peptides (CAMP) resistance to Salmonellae when it infects the host. pagP activated by PhoP- PhoQ help the bacteria from CAMP by increasing the acylation of Lipid A. Mutation in pagP gene elevates the leakage of outer membrane due to CAMP (Guo, et al., 1998).

Salmonella typhimurium PhoP and PhoQ system activates a gene named pagP necessary for the palmitate carrying Lipid A with hepta acylation that act as endotoxin antagonist and provides antibiotic resistance. pagP transfer palmitate to Lipid A precursor at n- linked hydroxy myristate. This is the first enzyme studied present in the outer membrane but plays a role in Lipid A synthesis. E. coli contain crcA homologous to pagP gene present in the outer membrane but required to synthesise Lipid A (Bishop et al., 2000).

PhoP and PhoQ are the genes that sense the outer environment and respond. These regulate the genes needed by bacteria for survival and antimicrobial resistance. PhoQ- PhoQ regulating system controls the alteration Lipid A structure by incorporation of aminoarabinose and two hydroxy myristate and altered Lipid A changes the expression of E- selectin from endothelial cells and tumour necrosis factor α expressed from monocytes and therefore help bacteria to survive in host tissue (Guo et al., 1997). A spontaneous mutation of pmrA gene in the Salmonella *typhimurium* enhances the resistance of bacteria against CAMP (Roland et al., 1993). Chemical analysis of LPS in pmrA (polymyxin resistant strain) mutant shows no change in the sugar, fatty acids and phosphate components with the mutant strain have 2- aminoethanol in excessive amount substitute the glycosidic pyrophosphate and form the diphosphate diester bond in approximately 40% Lipid A. Another unusual compound present in mutant strain is 4-amino-4-deoxy-L-arabinopyranose (L-Ara_p4N) attached to the Lipid A by 4- phosphate; otherwise, not present in the wild E. coli strain. These unusual amino arabinoses mediated esterification of Lipid A monophosphate and diphosphate provide the bacteria resistance against polymyxin (Nummila et al., 1995).

In *Leptospira interrogans* the structure of Lipid A was studied, which shows that Lpx genes that play a role in the biosynthesis of Lipid A have been homologous to the E. coli lpx genes but LPS of *L. Interrogans* activates TLR2 while LPS of E. coli activates TLR4. So, the structure of Lipid A was studied using different techniques report that Lipid A of L. *Interrogans*is hexa-acylated with the presence of 1,6-linked disaccharide derived from 2,3-diamino-2,3-dideoxy-D-glucopyranose. The disaccharide at 1- position have the phosphate group, which is methylated. Acylation of Lipid A at 3 and '3'with R-3-hydroxylaurate and at 2 and '2'with R-3-hydroxypalmitate done by LpxA. With this all at the distal residues of Lipid A two acyl chains are present. The responsive expression after the exposure of Lipid A isolated from *L. Interrogans* reveal that THP- 1 cell inactivates Lipid A while the RAW264.7 cells of mouse induce the expression of TNF (Que-Gewirth et al., 2004).

The respiratory tract of people suffering from cystic fibrosis has been infected with Pseudomonas aeruginosa. The structure of Lipid A of Pseudomonas aeruginosa in cystic fibrosis patients shows that Lipid A has palmitate and amino arabinose responsible for the resistance from CAMP and, therefore, increases the inflammatory reaction and develops respiratory diseases (Ernst et al., 1999). LPS of Pseudomonas aeruginosa also produces Lipid A having five or six acyl group depending on the environment.

Yersinia pestis, a human pathogen, synthesises several types of Lipid A depending upon the environment. At 37^{0}C Lipid A of *Yersinia pestis* is tetra-acylated, while at 26^{0}C or at low-temperature Lipid A is hexa- acylated. The study for the expression of genes responsible for the acylation shows that homologous genes *msbB* (*lpxM*) and *lpxP* encoding acyltransferase form hexa-acylated lipid IV_A by incorporation of C_{12} and $C_{16:1}$. The expression of hexa-acylated Lipid A is high in the flea midgut. Mutation in the *ΔmsbBΔlpxP* inhibits the formation of hexa acylated Lipid IV $_A$ become sensitive for cecropin (an antimicrobial peptide) while not to the antibiotic polymyxin B. The hexa- acylation of lipid IV A with C_{12} and $C_{16:1}$ modification at low temperature resembles the E. coli Lipid A at 12°C (Rebeil et al., 2006).

Table 3.1. Bacterial Lipid A modifications

Sr no	Bacteria	Change
1	*Chlamydia trachomatis*	Three KDO to Lipid A due to presence of gseA gene
2	*Francisella*	LpxD mofication form D1 and D2 D1: adds longer acyl chain with the 3-OH at acyl group of C18 of Lipid A (37^{0}C) D2: adds shorter acyl chain with the 3-OH at C16 of Lipid A (18^{0}C)
3	*Salmonella typhimurium*	pagP transfer palmitate to hepta acylated Lipid A' hyperacylation of Lipid A pmrA mutation: CAMP resistance
4	*Leptospirainterrogans*	Hexa- acylated with the presence of 1,6-linked disaccharide methylated phosphate at 1 position of disaccharide
5	*Pseudomonas aeruginosa*	Presence of palmitate and amino arabinose in Lipid A induces CAMP resistance
6	*Yersinia pestis*	Temperature dependent Lipid A modification 37^{0}C: tetra- acylated 26^{0}C or below: hexa- acylated

S. typhimurium has the LpxR gene expression that encodes for the enzyme 3′-O-deacylase. The LpxR have a unique hydrolase that removes Lipid 'A's 3′-acyloxyacyl residue. The LpxR orthologous are also present in E. coli strain, *Yersinia enterocolitica, Helicobacter pylori,* and *Vibrio cholerae*. Lpx may play a role in modifying cytokine expression by the animal infected with the

S. Typhimurium bacteria. So, this may conclude that this modification by LpxR in Lipid A structure due to deacylation help the bacteria to protect themselves from the innate immune system of the host during infection (Reynolds et al., 2006).

3.2. Core Oligosaccharide and O Antigen

The toxicity of bacteria depends on the Lipid A and the core region linked to Lipid A. Core region also has immunogenic effects (Holst, 2007). The core oligosaccharide has specificity and has immunogenicity. These branch frequently and have below fifteen sugars phosphorylated hetero-oligosaccharide, oligosaccharide which are highly conserved in the inner region and present near to the Lipid A (Silipo & Molinaro, 2010). These are also important for bacterial growth and outer membrane stability.

The core region of all bacteria contains 3-deoxy-D-*manno*-oct-2-ulopyranosonic acid. The variation in the core region is less than the O-specific polysaccharides (Unger, 1981). LPS core region of *Salmonella* in which at O-3 of second heptose L, D-Hep-(1→7)- L, D-Hep-(1→3)-L, D-Hep-(1→5)-Kdo is donated by glucopyranose (Glc*p). In the core region of Enterobacteria L-*glycero*-D-*manno*-heptopyranose (L,D-Hep) and the oligosaccharide L,D-Hep-(1→7)-L,D-Hep-(1→3)-L-α-D-Hep-(1→5)-[α-Kdo-(2→4)]-α-Kdo (Hep III, II, Hep, Kdo II and Kdo I) are present which induce the structural variations. With the L, D- Hep many LPS possess D-*glycero*-D-*manno*-heptopyranose (D, D-Hep) that is the precursor for L, D- Hep. Kdo II replacement in *Y. pestis* and *Serratia marcescens* with sugar D-*glycero*-D-*talo*-oct-2-ulopyranosonic acid (Ko) also results in structural variations, while in the Acinetobacter LPS Kdo I is changed (Gass et al., 1993, Kawahara et al., 1987 and Vinogradov et al., 1997).

Plesiomonas shigelloides have many serotypes such as O:54, O74, and O:13 Core region of *Plesiomonasshigelloides* do not have phosphate residues. The structure of the core region of serotypeO:13 and O:54 has a new property that is the presence of β -(1→4)-linked Gal residue to Hep I. In comparison, O:74 have GalA-(1→6)-β-Glc disaccharide at HepI. Linkage pattern for O antigen determined in all the three serotype shows that O:54 LPS is attached to branched oligosaccharides in place of O- 3 of HepII, while O:13 and O:74 both attached by same sugar but at various positions viz. galAto O-7 of HepIII and to the O-3 of Hep II. GalA is then replaced at O-2 with D, D-Hep (Lukasiewicz et al., 2006 and Niedziela *et al.*, 2006).

Yersinia pestis core region structure shows variations with temperature change, and therefore diversity is present in this bacterium. The structure of the core region of Y. Pestis is somewhat similar with the *Yersinia enterocolitica* having L,D-Hep-(1→7)-L,D-Hep-(1→3)-L,D-Hep-(1→5)-[Kdo-(2→4)-] Kdo positioned by β-Glc*N*Ac on O-3 of Hep II and on O-7 of Hep III by β-Gal or by D,D-Hep. The discussed substituents might change with the temperature change and with the species. At low temperature (6°C), the O-& of Ko or KDO II is replaced by 2-aminoethanol phosphate. This 2-aminoethanol phosphate was also found in the *enterica* and *E. coli* cultured at 37°C. The kO is present in this bacterium because Ko – (2→4) similar compound previously found in *Burkholderia* bacteria may replace KDO present in this bacteria (Holst, 2007).

Different serotypes of *Proteus penneri* studied for the core region of LPS. All the serotypes of this bacteria possess L, D-Hep-(1→7)-L, D-Hep-(1→3)-L, D-Hep-(1→5)-[Kdo-(2→4)-] Kdo carbohydrate backbone. The O-4 position of Hep I is occupied with L, D-Hep-(1→7)-L, D-Hep-(1→3)-L, D-Hep-(1→5)-[Kdo-(2→4)-] Kdo and GalA position at O-4 while 2-aminoethyl phosphate at O-6 of Hep II. This backbone of carbohydrate position differently and give structural specificity to the core (Vinogradov & Sidorczyk, 2002 and Vinogradov et al., 2002)

In K. pneumoniae three types of cores are found, and new one was found in 52145 strain (serotype O1: K2) (Vinogradov & Perry, 2001, Holst, 2002). The core region of this serotype does not have L, D-Hep-(1→4)-α-Kdo disaccharide at O- 6 of Glc*N* residue of the outer core but have β-Glc-(1→6)-Glc disaccharide at O-4 position of Glc*N* residue. However, this L, D-Hep-(1→4)-α-Kdo disaccharide is found in all the serotypes at O- 6 position. O-antigen is linked in the L, D-Hep-(1→4)-α-Kdo possessing serotype but at O-5 position of this KDO (Vinogradov et al., 2002b). Genes play a role in the formation of L, D-Hep-(1→4)-Kdo-(2→6)-Glc*N* sequence of outer core are WabH, WabI and WabJ (Frirdich et al., 2004). Where WabH transfers Glc*N*Ac to the GalA residue but in the outer structure Glc*N was found, and enzyme* WabN found that cause de-*N*-acetylation (Regué, 2005). GalA present in the outer membrane possess a negative charge and play a leading role in membrane integrity, like the phosphate present in other core types (Frirdich et al., 2005).

Five types of core structures R1, R2, R3, R4 and K-12, were studied thoroughly, where the structure of the inner core of R3 is very close to R2, and K-12 core type. R1, R3 and R4 *E. Coli* strains do not have waaZ gene. Study shows that changes in the inner core affect the outer core of LPS. waaZ gene encodes the change in *S. enterica* from Kdo III to Kdo II and R2 and K-12 in the case of E. coli when overexpressed results in the destruction of the outer core and therefore, reduces the O- antigen quantity (Muller-Loennies et al., 2002 and Frirdich et al., 2003). But another study report that waaZ gene expression was not elevated; therefore, it cannot be the cause for the destruction of the outer core, but its product is responsible for the change from Kdo III to Kdo II (Muller-Loennies, 2003).

The core types of LPS identified from E. coli are R1- R4 and the *S. enterica* have R1S and R2S (Müller-Loennies et al., 2003).

O- antigen polysaccharide is responsible for the structural variations. The sugar constitution and the association between monosaccharides are accountable for the O- antigen mediated structural variation. Each species has its different structure and chemical constitution, but all these factors also vary within the same species and same strain of bacteria. The modifications that occur in the O- antigen are such as alteration with the sugar moieties like residues of glucosyl and fucosyl non-stoichiometrically incorporation of non-carbohydrate compounds like acetyl or methyl groups. O- antigen modification may take part in the infection process from the adhesion of bacteria to the resistance from the host immune system. So, this may occur during the gene transfer, by the transformation of O- antigen, by alteration of O- antigen size or by fucosylation, glycosylation, acetylation. Now the question comes how this change in O- antigen occur?

The virulent strain *Salmonella enteritidis separated from the spleen of chick shows the high quantity of O- chain: core ratio in the LPS in contrast to a virulent strain of same bacteria* (Rahman et al., 1997).

In the enteric bacteria, wbcJ is the gene responsible for changing from GDP-D-mannose to GDP-L-fucose, where GDP-L-fucose is utilised for O-antigen synthesis. Bacteria strain mutant for this wbcJ gene does not have O-antigen and becomes sensitive to environmental stress (Aspinall et al., 1996, McGowan et al., 1998).

Symbiosis, any environmental condition, or infection that alters the LPS composition and O- antigen, also changes with the above conditions. During symbiosis, when the bacteria differentiate from bacterium to bacteroid the O-antigen composition varies may be in reaction to the alteration of environment, like decrease in oxygen is a reason for modification from water-loving (hydrophilic) to water hate (hydrophobic) form. In plants, the structure of bacterial O- antigen also alter during symbiosis due to the response given by the respective plant, which may be part of the symbiosis process. Any mutation in the O- chain inhibits symbiosis with the legume plant (Kannenberg & Carlson, 2001).

Conclusion

Endotoxins composed of Lipid A, oligosaccharides, and O- specific chain, where Lipid A is responsible for toxic activity of endotoxin and resistance of Gram-negative bacteria. Lipid A biosynthesis occurs in the interface between cytosol and the inner membrane. LpxA, B, C and D are the important enzymes involved in the synthesis of Lipid A. Other components of Lipid A such as oligosaccharides and O- specific chains also have immunogenic effects. But the above study shows that the compartment of toxicity depends not only on the Lipid A but also on other components such as O- antigen, and core structure of LPS also play a role in the resistance. So, the structure other than Lipid A can also be targeted to eliminate the antibiotic resistance problem.

References

Alshalchi, S. A., & Anderson, G. G. (2015). Expression of the lipopolysaccharide biosynthesis gene lpxD affects biofilm formation of Pseudomonas aeruginosa. *Archives of microbiology*, *197*(2), 135-145.

Anderson, M. S., Bulawa, C. E., & Raetz, C. R. (1985). The biosynthesis of gram-negative endotoxin. Formation of Lipid A precursors from UDP-GlcNAc in extracts of Escherichia coli. *Journal of Biological Chemistry*, *260*(29), 15536-15541.

Anderson, M. S., Bull, H. G., Galloway, S. M., Kelly, T. M., Mohan, S. A. N. D. H. Y. A., Radika, K., & Raetz, C. R. (1993). UDP-N-acetylglucosamine acyltransferase of Escherichia coli. The first step of endotoxin biosynthesis is thermodynamically unfavorable. *Journal of Biological Chemistry*, *268*(26), 19858-19865.

Aronson, A. (2002). Sporulation and δ-endotoxin synthesis by Bacillus thuringiensis. *Cellular and Molecular Life Sciences CMLS*, *59*(3), 417-425.

Aspinall, G. O., Monteiro, M. A., Pang, H., Walsh, E. J., & Moran, A. P. (1996). Lipopolysaccharide of the Helicobacter pylori type strain NCTC 11637 (ATCC 43504): structure of the O antigen chain and core oligosaccharide regions. *Biochemistry*, *35*(7), 2489-2497.

Babinski, K. J., Kanjilal, S. J., & Raetz, C. R. (2002). Accumulation of the Lipid A precursor UDP-2, 3-diacylglucosamine in an Escherichia coli mutant lacking the lpxH gene. *Journal of Biological Chemistry*, *277*(29), 25947-25956.

Babinski, K. J., Ribeiro, A. A., & Raetz, C. R. (2002). The Escherichia coli gene encoding the UDP-2, 3-diacylglucosamine pyrophosphatase of Lipid A biosynthesis. *Journal of Biological Chemistry*, *277*(29), 25937-25946.

Bartling, C. M., & Raetz, C. R. (2008). Steady-state kinetics and mechanism of LpxD, the N-acyltransferase of Lipid A biosynthesis. *Biochemistry*, *47*(19), 5290-5302.

Belunis, C. J., Mdluli, K. E., Raetz, C. R., & Nano, F. E. (1992). A novel 3-deoxy-D-manno-octulosonic acid transferase from Chlamydia trachomatis required for expression of the genus-specific epitope. *Journal of Biological Chemistry*, *267*(26), 18702-18707.

Bishop, R. E., Gibbons, H. S., Guina, T., Trent, M. S., Miller, S. I., & Raetz, C. R. (2000). Transfer of palmitate from phospholipids to lipid A in outer membranes of Gram-negative bacteria. *The EMBO journal*, *19*(19), 5071-5080.

Bohl, T. E., Shi, K., Lee, J. K., & Aihara, H. (2018). Crystal structure of Lipid A disaccharide synthase LpxB from Escherichia coli. *Nature communications*, *9*(1), 1-13.

Brozek, K. A., & Raetz, C. R. (1990). Biosynthesis of lipid A in Escherichia coli. Acyl carrier protein-dependent incorporation of laurate and myristate. *Journal of Biological Chemistry*, *265*(26), 15410-15417.

Buetow, L., Smith, T. K., Dawson, A., Fyffe, S., & Hunter, W. N. (2007). Structure and reactivity of LpxD, the N-acyltransferase of Lipid A biosynthesis. *Proceedings of the National Academy of Sciences*, *104*(11), 4321-4326.

Clayton, G. M., Klein, D. J., Rickert, K. W., Patel, S. B., Kornienko, M., Zugay-Murphy, J., ... & Soisson, S. M. (2013). Structure of the bacterial deacetylase LpxC bound to the nucleotide reaction product reveals mechanisms of oxyanion sstabilisationand proton transfer. *Journal of Biological Chemistry*, *288*(47), 34073-34080.

Clementz, T., & Raetz, C. R. (1991). A gene coding for 3-deoxy-D-manno-octulosonic-acid transferase in Escherichia coli. Identification, mapping, cloning, and sequencing. *Journal of Biological Chemistry*, *266*(15), 9687-9696.

Clementz, T., Bednarski, J. J., & Raetz, C. R. (1996). Function of the htrB High Temperature Requirement Gene of Escherichia coli in the Acylation of Lipid A: HtrB CATALYZED INCORPORATION OF LAURATE (∗). *Journal of Biological Chemistry*, *271*(20), 12095-12102.

Clementz, T., Zhou, Z., & Raetz, C. H. (1997). Function of the Escherichia coli msbB gene, a multicopy suppressor of htrB knockouts, in the acylation of lipid A: acylation by MsbB follows laurate incorporation by HtrB. *Journal of Biological Chemistry*, *272*(16), 10353-10360.

Corsaro, M. M., Pieretti, G., Lindner, B., Lanzetta, R., Parrilli, E., Tutino, M. L., & Parrilli, M. (2008). Highly phosphorylated core oligosaccaride structures from cold-adapted Psychromonasarctica. *Chemistry–A European Journal*, *14*(30), 9368-9376.

Crowell, D. N., Anderson, M. S., & Raetz, C. R. (1986). Molecular cloning of the genes for Lipid A disaccharide synthase and UDP-N-acetylglucosamine acyltransferase in Escherichia coli. *Journal of bacteriology*, *168*(1), 152-159.

Emptage, R. P., Pemble IV, C. W., York, J. D., Raetz, C. R., & Zhou, P. (2013). Mechanistic scharacterisationof the tetraacyldisaccharide-1-phosphate 4′-kinase LpxK involved in Lipid A biosynthesis. *Biochemistry*, *52*(13), 2280-2290.

Ernst, R. K., Eugene, C. Y., Guo, L., Lim, K. B., Burns, J. L., Hackett, M., & Miller, S. I. (1999). Specific lipopolysaccharide found in cystic fibrosis airway Pseudomonas aeruginosa. *Science*, *286*(5444), 1561-1565.

Frirdich, E., Bouwman, C., Vinogradov, E., & Whitfield, C. (2005). The role of galacturonic acid in outer membrane stability in Klebsiella pneumoniae. *Journal of Biological Chemistry*, *280*(30), 27604-27612.

Frirdich, E., Lindner, B., Holst, O., & Whitfield, C. (2003). Overexpression of the waaZ gene leads to modification of the structure of the inner core region of Escherichia coli lipopolysaccharide, truncation of the outer core, and reduction of the amount of O polysaccharide on the cell surface. *Journal of bacteriology*, *185*(5), 1659-1671.

Frirdich, E., Vinogradov, E., & Whitfield, C. (2004). Biosynthesis of a novel 3-deoxy-D-manno-oct-2-ulosonic acid-containing outer core oligosaccharide in the lipopolysaccharide of Klebsiella pneumoniae. *Journal of Biological Chemistry*, *279*(27), 27928-27940.

Garrett, T. A., Kadrmas, J. L., & Raetz, C. R. (1997). Identification of the gene encoding the Escherichia coli lipid A 4′-kinase: facile phosphorylation of endotoxin analogs with recombinant LpxK. *Journal of Biological Chemistry*, *272*(35), 21855-21864.

Gass, J., Strobl, M., Loibner, A., Kosma, P., & Zähringer, U. (1993). Synthesis of allyl O-[sodium (α-d-glcero-d-talo-2-octulopyranosyl) onate]-(2→6)-2-acetamido-2-deoxy-β-d-glucopyranoside, a core constituent of the lipopolysac-charide from Acinetobacter calcoaceticus NCTC 10305. *Carbohydrate research*, *244*(1), 69-84.

Gattis, S. G., Chung, H. S., Trent, M. S., & Raetz, C. R. (2013). The origin of 8-amino-3, 8-dideoxy-D-manno-octulosonic acid (Kdo8N) in the lipopolysaccharide of Shewanellaoneidensis. *Journal of Biological Chemistry*, *288*(13), 9216-9225.

Gmeiner, J., Simon, M., & LÜDERITZ, O. (1971). The Linkage of Phosphate Groups and of 2-Keto-3-deoxyoctonate to the Lipid A Component in a Salmonella minnesota Lipopolysaccharide. *European journal of biochemistry*, *21*(3), 355-356.

Goldman, R. C., & Devine, E. M. (1987). Isolation of Salmonella typhimurium strains that sutiliseexogenous 3-deoxy-D-manno-octulosonate for synthesis of lipopolysaccharide. *Journal of bacteriology*, *169*(11), 5060-5065.

Goldman, R. C., Bolling, T. J., Kohlbrenner, W. E., Kim, Y., & Fox, J. L. (1986). Primary structure of CTP: CMP-3-deoxy-D-manno-octulosonate cytidylyltrans-ferase (CMP-KDO synthetase) from Escherichia coli. *Journal of Biological Chemistry*, *261*(34), 15831-15835.

Gorbet, M. B., & Sefton, M. V. (2005). Endotoxin: the uninvited guest. *Biomaterials*, *26*(34), 6811-6817.

Guo, L., Lim, K. B., Gunn, J. S., Bainbridge, B., Darveau, R. P., Hackett, M., & Miller, S. I. (1997). Regulation of Lipid A modifications by Salmonella typhimurium virulence genes phoP-phoQ. *Science*, *276*(5310), 250-253.

Guo, L., Lim, K. B., Poduje, C. M., Daniel, M., Gunn, J. S., Hackett, M., & Miller, S. I. (1998). Lipid A acylation and bacterial resistance against vertebrate antimicrobial peptides. *Cell*, *95*(2), 189-198.

Hankins, J. V., & Trent, M. S. (2009). Secondary acylation of Vibrio cholerae lipopolysaccharide requires phosphorylation of Kdo. *Journal of Biological Chemistry*, *284*(38), 25804-25812.

Hankins, J. V., & Trent, M. S. (2009). Secondary acylation of Vibrio cholerae lipopolysaccharide requires phosphorylation of Kdo. *Journal of Biological Chemistry*, *284*(38), 25804-25812.

HELANDER, I. M., LINDNER, B., BRADE, H., ALTMANN, K., LINDBERG, A. A., RIETSCHEL, E. T., & ZÄHRINGER, U. (1988). Chemical structure of the lipopolysaccharide of Haemophilus influenzae strain I-69 Rd−/b+ Description of a novel deep-rough chemotype. *European journal of biochemistry*, *177*(3), 483-492.

Holst, O. (2007). The structures of core regions from enterobacterial lipopolysaccharides–an update. *FEMS microbiology letters*, *271*(1), 3-11.

Kannenberg, E. L., & Carlson, R. W. (2001). Lipid A and O-chain modifications cause Rhizobium lipopolysaccharides to become hydrophobic during bacteroid development. *Molecular microbiology*, *39*(2), 379-392.

Kawahara, K., Brade, H., Rietschel, E. T., & Zähringer, U. (1987). Studies on the chemical structure of the core-lipid A region of the lipopolysaccharide of Acinetobacter calcoaceticus NCTC 10305: Detection of a new 2-octulosonic acid interlinking the core oligosaccharide and lipid A component. *European journal of biochemistry*, *163*(3), 489-495.

Li, Y., Powell, D. A., Shaffer, S. A., Rasko, D. A., Pelletier, M. R., Leszyk, J. D., ... & Ernst, R. K. (2012). LPS remodeling is an evolved survival strategy for bacteria. *Proceedings of the National Academy of Sciences*, *109*(22), 8716-8721.

Lukasiewicz, J., Dzieciatkowska, M., Niedziela, T., Jachymek, W., Augustyniuk, A., Kenne, L., & Lugowski, C. (2006). Complete lipopolysaccharide of Plesiomonasshigelloides O74: H5 (strain CNCTC 144/92). 2. Lipid A, its structural variability, the linkage to the core oligosaccharide, and the biological activity of the lipopolysaccharide. *Biochemistry*, *45*(35), 10434-10447.

McGowan, C. C., Necheva, A., Thompson, S. A., Cover, T. L., & Blaser, M. J. (1998). Acid-induced expression of an LPS-associated gene in Helicobacter pylori. *Molecular microbiology*, *30*(1), 19-31.

Metzger IV, L. E., & Raetz, C. R. (2010). An alternative route for UDP-diacylglucosamine hydrolysis in bacterial Lipid A biosynthesis. *Biochemistry*, *49*(31), 6715-6726.

Miller, S. I., Kukral, A. M., & Mekalanos, J. J. (1989). A two-component regulatory system (phoPphoQ) controls Salmonella typhimurium virulence. *Proceedings of the National Academy of Sciences*, *86*(13), 5054-5058.

Mohan, S., & Raetz, C. R. (1994). Endotoxin biosynthesis in Pseudomonas aeruginosa: enzymatic incorporation of laurate before 3-deoxy-D-manno-octulosonate. *Journal of bacteriology*, *176*(22), 6944-6951.

Moran, A. P., Zähringer, U., Seydel, U., Scholz, D., Stütz, P., & Rietschel, E. T. (1991). Structural analysis of the Lipid A component of Campylobacter jejuni CCUG 10936 (serotype O: 2) lipopolysaccharide: Description of a lipid A containing a hybrid backbone of 2-amino-2-deoxy-D-glucose and 2, 3-diamino-2, 3-dideoxy-D-glucose. *European journal of biochemistry*, *198*(2), 459-469.

Müller-Loennies, S., Brade, L., MacKenzie, C. R., Di Padova, F. E., & Brade, H. (2003). Identification of a cross-reactive epitope widely present in lipopolysaccharide from enterobacteria and srecognisedby the cross-protective monoclonal antibody WN1 222-5. *Journal of Biological Chemistry*, *278*(28), 25618-25627.

Müller-Loennies, S., Lindner, B., & Brade, H. (2002). Structural analysis of deacylated lipopolysaccharide of Escherichia coli strains 2513 (R4 core-type) and F653 (R3 core-type). *European journal of biochemistry*, *269*(23), 5982-5991.

Needham, B. D., & Trent, M. S. (2013). Fortifying the barrier: the impact of Lipid A remodelling on bacterial pathogenesis. *Nature Reviews Microbiology*, *11*(7), 467-481.

Niedziela, T., Dag, S., Lukasiewicz, J., Dzieciatkowska, M., Jachymek, W., Lugowski, C., & Kenne, L. (2006). Complete lipopolysaccharide of Plesiomonasshigelloides O74: H5 (strain CNCTC 144/92). 1. Structural analysis of the highly hydrophobic lipopolysaccharide, including the O-antigen, its biological repeating unit, the core oligosaccharide, and the linkage between them. *Biochemistry*, *45*(35), 10422-10433.

Nummila, K., Kilpeläinen, I., Zähringer, U., Vaara, M., & Helander, I. M. (1995). Lipopolysaccharides of polymyxin B-resistant mutants of Escherichia coii are extensively substituted by 2-aminoethyl pyrophosphate and contain aminoarabinose in lipid A. *Molecular microbiology*, *16*(2), 271-278.

Prost, L. R., & Miller, S. I. (2008). The Salmonellae PhoQ sensor: mechanisms of detection of phagosome signals. *Cellular microbiology*, *10*(3), 576-582.

Que-Gewirth, N. L., Ribeiro, A. A., Kalb, S. R., Cotter, R. J., Bulach, D. M., Adler, B., ... & Raetz, C. R. (2004). A methylated phosphate group and four amide-linked acyl chains in Leptospira interrogans lipid A: the membrane anchor of an unusual lipopolysaccharide that activates TLR2. *Journal of Biological Chemistry*, *279*(24), 25420-25429.

Raetz, C. R. (1993). Bacterial endotoxins: extraordinary lipids that activate eucaryotic signal transduction. *Journal of bacteriology*, *175*(18), 5745-5753.

Raetz, C. R., & Whitfield, C. (2002). Lipopolysaccharide endotoxins. *Annual review of biochemistry*, *71*(1), 635-700.

Raetz, C. R., Reynolds, C. M., Trent, M. S., & Bishop, R. E. (2007). Lipid A modification systems in gram-negative bacteria. *Annu. Rev. Biochem.*, *76*, 295-329.

Rahman, M. M., Guard-Petter, J., & Carlson, R. W. (1997). A virulent isolate of Salmonella enteritidis produces a Salmonella typhi-like lipopolysaccharide. *Journal of bacteriology*, *179*(7), 2126-2131.

Rebeil, R., Ernst, R. K., Jarrett, C. O., Adams, K. N., Miller, S. I., & Hinnebusch, B. J. (2006). sCharacterisationof late acyltransferase genes of Yersinia pestis and their role in temperature-dependent Lipid A variation. *Journal of bacteriology*, *188*(4), 1381-1388.

Regué, M., Izquierdo, L., Fresno, S., Jimenez, N., Piqué, N., Corsaro, M. M., ... & Tomás, J. M. (2005). The incorporation of glucosamine into enterobacterial core lipopolysaccharide: two enzymatic steps are required. *Journal of Biological Chemistry*, *280*(44), 36648-36656.

Reynolds, C. M., Ribeiro, A. A., McGrath, S. C., Cotter, R. J., Raetz, C. R., & Trent, M. S. (2006). An outer membrane enzyme encoded by Salmonella typhimurium lpxR that removes the 3′-acyloxyacyl moiety of lipid A. *Journal of Biological Chemistry*, *281*(31), 21974-21987.

Rick, P. D., Fung, L. W., Ho, C., & Osborn, M. J. (1977). Lipid A mutants of Salmonella typhimurium. Purification and scharacterisationof a lipid A precursor produced by a mutant in 3-deoxy-D-mannooctulosonate-8-phosphate synthetase. *Journal of Biological Chemistry*, *252*(14), 4904-4912.

Rietschel, E. T., Kirikae, T., Schade, F. U., Mamat, U., Schmidt, G., Loppnow, H., ... & Brade, H. (1994). Bacterial endotoxin: molecular relationships of structure to activity and function. *The FASEB Journal*, *8*(2), 217-225.

Rietschel, E. T., Wollenweber, H. W., Russa, R., Brade, H., & Zähringer, U. (1984). Concepts of the chemical structure of lipid A. *Clinical Infectious Diseases*, *6*(4), 432-438.

Roland, K. L., Martin, L. E., Esther, C. R., & Spitznagel, J. K. (1993). Spontaneous pmrA mutants of Salmonella typhimurium LT2 define a new two-component regulatory system with a possible role in virulence. *Journal of bacteriology*, *175*(13), 4154-4164.

Rosenfeld, Y., & Shai, Y. (2006). Lipopolysaccharide (Endotoxin)-host defense antibacterial peptides interactions: role in bacterial resistance and prevention of sepsis. *Biochimica et Biophysica Acta (BBA)-Biomembranes*, *1758*(9), 1513-1522.

Sampath, V. (2018). Bacterial endotoxin-lipopolysaccharide; structure, function and its role in immunity in vertebrates and invertebrates. *Agriculture and Natural Resources*, *52*(2), 115-120.

Silipo, A., & Molinaro, A. (2010). The diversity of the core oligosaccharide in lipopolysaccharides. In *Endotoxins: Structure, Function and Recognition* (pp. 69-99). Springer, Dordrecht.

Tomaras, A. P., McPherson, C. J., Kuhn, M., Carifa, A., Mullins, L., George, D., ... & O'Donnell, J. P. (2014). LpxC inhibitors as new antibacterial agents and tools for studying regulation of Lipid A biosynthesis in Gram-negative pathogens. *MBio*, *5*(5), e01551-14.

Trent, M. S., Stead, C. M., Tran, A. X., & Hankins, J. V. (2006). Invited review: diversity of endotoxin and its impact on pathogenesis. *Journal of endotoxin research*, *12*(4), 205-223.

Unger, F. M. (1981). The chemistry and biological significance of 3-deoxy-D-manno-2-octulosonic acid (KDO). In *Advances in carbohydrate chemistry and biochemistry* (Vol. 38, pp. 323-388). Academic Press.

Vinogradov, E. V., Müller-Loennies, S., Petersen, B. O., Meshkov, S., Thomas-Oates, J. E., Holst, O., & Brade, H. (1997). Structural investigation of the lipopolysaccharide from Acinetobacter haemolyticus strain NCTC 10305 (ATCC 17906, DNA group 4). *European journal of biochemistry*, *247*(1), 82-90.

Vinogradov, E., & Perry, M. B. (2001). Structural analysis of the core region of the lipopolysaccharides from eight serotypes of Klebsiella pneumoniae. *Carbohydrate Research*, *335*(4), 291-296.

Vinogradov, E., & Sidorczyk, Z. (2002). The structure of the carbohydrate backbone of the rough type lipopolysaccharides from Proteus penneri strains 12, 13, 37 and 44. *Carbohydrate research*, *337*(9), 835-840.

Vinogradov, E., Frirdich, E., MacLean, L. L., Perry, M. B., Petersen, B. O., Duus, J. Ø., & Whitfield, C. (2002). Structures of lipopolysaccharides from Klebsiella pneumoniae: Elucidation of the structure of the linkage region between core and polysaccharide O chain and identification of the residues at the non-reducing termini of the Ochains. *Journal of Biological Chemistry*, *277*(28), 25070-25081.

White, K. A., Kaltashov, I. A., Cotter, R. J., & Raetz, C. R. (1997). A mono-functional 3-deoxy-D-manno-octulosonic acid (Kdo) transferase and a Kdo kinase in extracts of Haemophilus influenzae. *Journal of Biological Chemistry*, *272*(26), 16555-16563.

Whitfield, C., & Trent, M. S. (2014). Biosynthesis and export of bacterial lipopolysaccharides. *Annual review of biochemistry*, *83*, 99-128.

Whitfield, C., Amor, P. A., & Koplin, R. (1997). Modulation of the surface architecture of Gram-negative bacteria by the action of surface polymer: Lipid A–core ligase and by determinants of polymer chain length. *Molecular microbiology*, *23*(4), 629-638.

Whittington, D. A., Rusche, K. M., Shin, H., Fierke, C. A., & Christianson, D. W. (2003). Crystal structure of LpxC, a zinc-dependent deacetylase essential for endotoxin biosynthesis. *Proceedings of the National Academy of Sciences*, *100*(14), 8146-8150.

Wyckoff, T. J., Raetz, C. R., & Jackman, J. E. (1998). Antibacterial and anti-inflammatory agents that target endotoxin. *Trends in microbiology*, *6*(4), 154-159.

Zhou, P., & Barb, A. W. (2008). Mechanism and inhibition of LpxC: an essential zinc-dependent deacetylase of bacterial Lipid A synthesis. *Current pharmaceutical biotechnology*, *9*(1), 9-15.

Chapter 4

Endotoxins: Regulation of the Virulence

Shubhanshi Ranjan[1,*] and Gurvinder Kaur[2]

[1]Masters Scholar, College of Animal Biotechnology,
Guru Angad Dev Veterinary and Animal Sciences University, Ludhiana, Punjab, India
[2]Teaching Assistant, College of Animal Biotechnology,
Guru Angad Dev Veterinary and Animal Sciences University, Ludhiana, Punjab, India

Abstract

The type of condition that develops and the risk associated with it can be influenced by the timing of an agent's exposure during one's lifetime. Cholera, plague, whooping cough, and certain type of meningitis are all caused by endotoxins, which are gram-negative bacteria molecules. Pathogenesis is caused by endotoxins entering the host cell. Endotoxins are not heat-labile, but they can be neutralised by oxidising agents such as superoxide, peroxide, and hypochlorites. It's important to remember that LPS is found on the bacterium's outermost surface when interacting with different hosts. A host's susceptibility to any xenobiotic or compound varies. The age, family history, and immune function of an infection-prone host all play a role. Bacteria that infect their hosts by invading their cells are classified as either obligate or facultative intracellular bacteria. The main causes of invasion are unknown, but it appears that a number of gene products are involved. Pathogens can bypass the innate immune system and reach distant organs thanks to bacterial toxins produced by gram-negative bacteria. The type of toxin has an impact on the disease's outcome. In the last decade, the mechanisms by which these toxins modulate host defence have become much clearer. This could result in more accurate toxin targeting and improved clinical outcomes.

* Corresponding Author's Email: r.shubhanshi109@gmail.com.

In: Endotoxins and their Importance
Editors: Arif Pandit and R. S. Sethi
ISBN: 978-1-68507-839-3

Keywords: toxin, LPS, host, immune system

Introduction

During one's lifetime, the timing of exposure to an agent can influence the specific type of condition that may develop as well as the level of risk associated with that exposure. Endotoxins are potentially lethal molecules produced by many gram-negative bacteria, such as those that cause cholera, plague, whooping cough, and a type of meningitis, and they continue to captivate biological researchers and others in their pursuit of better understanding. Endotoxin is sometimes referred to as any cell-associated bacterial toxin, in other words, it is referred to as a lipopolysaccharide complex that is associated with the outer membrane of Gram-negative pathogens such as *Escherichia coli, Salmonella, Shigella, Pseudomonas, Neisseria, Haemophilus influenzae, Bordetella pertussis,* and *Vibrio cholerae*. It takes a variety of factors for endotoxins to enter the host cell and for pathogenesis to develop, and this is a complex process that requires a variety of factors.

4.1. Host Susceptibility

Endotoxins are not heat-labile, which means that they do not become destabilised when heated; however, they can be neutralised by specific oxidising agents such as superoxide, peroxide, and hypochlorite, among others. Furthermore, the location of the LPS on the bacterium's outermost surface provides it with the greatest opportunity to interact with the various hosts, and in many cases, it is the host's response to the gram-negative organism, rather than the organism itself, that poses the greatest threat to the host's tissue. A variety of factors influence the susceptibility of a host to any xenobiotic or compound that is introduced into it. Age, family history, and immune function are all important factors in determining a host's susceptibility to any type of infection.

4.2. Factors Enhancing the Host Susceptibility to Endotoxins

The pyrogenic exotoxin (PE) type C produced by *Staphylococcus aureus* can increase the susceptibility of rabbits to lethal shock by endotoxin by as much as 50,000-fold and 40,000-fold, respectively. In addition to causing fever, this toxin has the ability to activate the immune system, which includes non-specific stimulation of T lymphocytes, suppression of immunoglobulin M synthesis, and the enhancement of hypersensitivity, which can result in the development of a skin rash after being exposed to endotoxin (Patrick, 1982). To overcome this shock, one can be immunised with pyrogenic exotoxin (PE) type C toxin in order to be protected from increased susceptibility to endotoxin.

Endotoxins can also be affected by pre-existing infections or intestinal microbial flora, which can alter their action. It is possible to increase the lethality of endotoxin in animals that have previously been injected with BCG, *Brucella, Hemophilus pertussis*, and a variety of lymphocytic viruses. An important source of LPS for vertebrates is their resident gut microbiota, and intestinal alkaline phosphatase helps to detoxify the LPS and prevent intestinal inflammation as a result of the resident microbiota (Bates et al., 2007)

Arachidonate metabolites such as Leukotrienes and other cysteinyl leukotrienes have been implicated as mediators of endotoxin action in the past. They identify mononuclear phagocytes (macrophages) as the target cells of LPS's *in vivo* action, and they direct their attention to them. Shock-like reactions, myocardial depression, and plasma extravasation have all been reported following the injection of these mediators into sensitive species. Endotoxin shock in sensitive species is illustrated in Figure 1 by the generalised effect of arachidonate mediators such as cysteinyl leukotrienes, thromboxane, and prostaglandins on increasing the susceptibility to endotoxin shock.

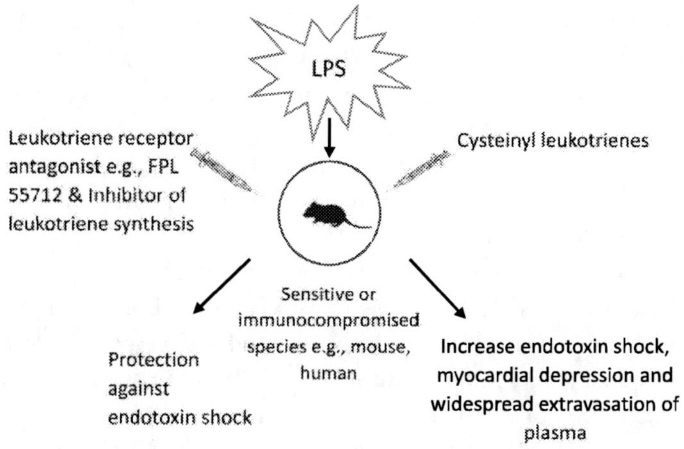

When different antimicrobial therapies are administered, they have the potential to cause a variety of diseased conditions, such as sepsis. Endotoxin exposure can result in the development of respiratory distress syndrome and disseminated intravascular coagulation in these circumstances. Occasionally, antibiotic therapy can result in the release of bacterial endotoxin-like products, which can result in a Jarisch–Herxheimer reaction in the case of bacteremia infections. Ji and colleagues (1993).

4.3. LPS Immunogenicity

It is composed of three components or regions: lipid A, a R polysaccharide, and an O polysaccharide. Lipid A is the most abundant component. The O polysaccharide in the Gram-negative cell wall contains the most important antigenic determinant (antibody-combining site) in the cell wall. Lipid A is composed of a phosphorylated N-acetylglucosamine (NAG) dimer with 6 or 7 fatty acids (FA) attached to it. Lipid B is composed of a phosphorylated N-acetylglucosamine (NAG) dimer with 6 or 7 fatty acids (FA) attached to it. NAGs are attached to the six positions of one NAG by the core (R) antigen or the R polysaccharide. The R antigen consists of a short chain of sugars that binds to the receptor. For example, KDO - Hep-Hep - Glu - Gal - Glu - GluNAc - Somatic (O) antigen or O polysaccharide is attached to the core polysaccharide, while KDO - Hep-Hep - Glu - Gal - Glu - GluNAc - O polysaccharide is attached to the core polysaccharide. It is composed of repeating oligosaccharide subunits, each of which contains three to five sugars. The composition of the sugars in the O side chain varies significantly between species and even between strains of Gram-negative bacteria, and this variation is particularly pronounced. Variations in the sugar content of the O polysaccharide contribute to the wide variety of antigenic types found in *Salmonella* and *E. coli*, as well as other strains of Gram-negative species, according to some researchers. In addition to the "smoothness" (colony morphology) of the strain, certain sugars in the structure, particularly the terminal ones, confer immunological specificity to the O antigen. The loss of the O specific region due to mutation results in the strain developing a "rough" colony morphology, which is referred to as a R strain. Loss of the O antigen results in a partial loss of virulence in *E. coli* and *Salmonella*, indicating that this portion of LPS is required during a host-parasite interaction in these bacteria. Such "rough" mutants are known to be more susceptible to phagocytosis and serum bactericidal reactions than their "clean" counterparts.

It is believed that both Lipid A (the toxic component of LPS) and the polysaccharide side chains (the nontoxic but immunogenic component of LPS) are important factors in the development of virulence in Gram-negative bacteria.

4.4. The O Polysaccharide and Virulence

Virulence, or the presence of the major antigenic determinant site and the property of "smoothness," is associated with the presence of an intact O polysaccharide. In the O polysaccharide, even a small change in the sugar sequences of the side chains leads to an increase in the virulence property of the polysaccharide. Demonstrating that O polysaccharides play a role in virulence induction can be done in a variety of ways, including demonstrating the ability of organisms to adhere precisely to specific tissues, such as epithelial tissues, and demonstrating the hydrophilic O polysaccharides' ability to significantly affect the water binding capacity of cells and act as water-soluble carriers for toxic Lipid A. Additionally, smooth antigens may protect against antibody and complement activation-mediated toxicity, as well as serving as the basis for antigenic variation among various gram-negative pathogens such as *E. coli, Salmonella,* and *Vibrio cholerae*. Smooth antigens have been shown to confer resistance to phagocytes and to protect against antibody and complement activation-mediated toxicity in animal models. It has multiple opportunities to infect its host because of the bacterium's serological properties, which allows it to circumvent the immune response against a different serotype of bacteria. Following a thorough review of the literature, it appears that O polysaccharides can only rarely elicit memory immune responses that protect the host from secondary exposure to a specific endotoxin.

4.4.1. Lipid A and Virulence

The endotoxin lipid, which is a component of the endotoxin, is critical in the induction of the mammalian immune system. With long-chain hydrocarbons and amide linked fatty acids that have been -OH substituted, it contains the ester-linked hydrophobic, membrane-anchoring region of LPS as well as other components. A large number of biological activities associated with endotoxins from bacteria are therefore associated with this chemically unique amphipathic structure. This lipid A portion is where the endotoxin-associated proteins are primarily found, and they are referred to as "endotoxin protein" (EP)2 or "lipid A-associated protein" (LAP).

The injection of live or killed Gram-negative cells or purified LPS into experimental animals results in a wide range of non-specific reactions such as fever, increased infection, inflammatory reactions, disseminated intravascular coagulation, hypotension, shock, and death, amongst other symptoms. The injection of relatively small doses of endotoxin results in the death of the vast majority of mammalian organisms tested. These are the events that occur in a predictable sequence: (1) the latent period; (2) physiological distress (diarrhoea, prostration, shock); and (3) death. The amount of time it takes for an animal to die is dependent on the dose of endotoxin used, the route of administration used, and the species of animal used, each of which has a different level of susceptibility.

Multiple organ failure and lethality can result from a systemic hyper-inflammatory response induced by LPS. IL-8, IL-6, IL-1a, IL-1b, IL-12, and IFN are all produced by inflammatory cells as a result of LPS stimulation of the immune system. During an endotoxic shock episode, TNF plays a crucial role. Acute exposure to endotoxin can result in life-threatening sepsis, whereas chronic exposure has been linked to a variety of disease states affecting the gastrointestinal, nervous, metabolic, vascular, pulmonary, and immune systems, among other organs. (Champion et al., 2013). The mechanism of action of LPS is complex, which involves the interaction of pathogen-associated molecular patterns of LPS with lipid-binding protein (LBP) in the serum, CD14 on the cell membrane and MD2, which associates with Toll-like receptor-4 (TLR4) creating CD14/TLR4/LBP complex present on major immune cells like macrophages and dendritic cells. Further, this signal is passed to IL-1 receptor-associated kinases (IRAKs). TNF associated factor 6 (TRAF6), and NF-kappaB inducing kinase resulting in activation of NF-kappaB (Palsson and ONeill., 2004). In monocytes and macrophages, three types of events are triggered during their interaction with LPS:

1. **Production of cytokines**, including IL-1, IL-6, IL-8, tumour necrosis factor (TNF) and platelet-activating factor. These, in turn, stimulate the production of prostaglandins and leukotrienes. These are potent mediators of inflammation and septic shock that accompanies endotoxin toxaemia. LPS activates macrophages to enhance phagocytosis and cytotoxicity. Macrophages are stimulated to produce and release lysosomal enzymes, IL-1 ("endogenous

pyrogen"), tumour necrosis factor (TNFalpha), as well as other cytokines and mediators.

2. **Activation of the complement cascade**. C3a and C5a cause histamine release (leading to vasodilation) and affect neutrophil chemotaxis and accumulation. The result is inflammation.
3. **Activation of the coagulation cascade**. Initial activation of Hageman factor (blood clotting Factor XII) can activate several humoral systems resulting in
   (a) coagulation: a blood clotting cascade that leads to coagulation, thrombosis, acute disseminated intravascular coagulation, which depletes platelets and various clotting factors resulting in internal bleeding.
   (b) activation of the complement alternative pathway (as above, which leads to inflammation)
   (c) plasmin activation, which leads to fibrinolysis and haemorrhaging.
   (d) kinin activation releases bradykinins and other vasoactive peptides, which causes hypotension.

The net effect induces inflammation, intravascular coagulation, haemorrhage and shock.

LPS also acts as a **B cell mitogen**, stimulating the polyclonal differentiation and multiplication of B-cells and the secretion of immunoglobulins, especially IgG and IgM.

4.5. Host Resistance

The host has its defence mechanisms to resist any foreign body invasion to its system, and the pathogen has to defeat these barriers to infect the organism successfully. The host can have specific direct antimicrobial activity or inhibition of pathogen attachment by protection with skin, certain degenerative enzymes and barriers with the mucus secretion. If the pathogen crosses such primary barriers, it has to face the other levels of the host resistance, i.e., the non-specific or specific immune resistance mechanisms.

The host immune system, composed of different immune cells, is primarily programmed to identify and destroy or neutralize the subject pathogen. Nevertheless, as the organism ages, its immune response weakens as a generalized decline is seen, and therefore with age, the host gets more susceptible to infection with the pathogen. This is more precisely termed age-associated immune deficiency. Likewise, humans have numerous mechanisms to inactivate Lipopolysaccharide toxicity, such as lipid A neutralizing proteins (lactoferrin, collectins, lysozyme, etc.), specific and cross-reactive anti-LPS antibodies sequestration of the lipid A moiety within lipoprotein micelles. Although the total activity of the immune system is not halted, total antibody production is unchanged, but the response towards a foreign antigen becomes less. This is probably due to the increased production of antibodies against certain self-antigens. Similarly, even the production of specific cytokines is increased e.g., the interleukin-4 (IL-4), while certain of them even decrease in production like interleukin-2 (IL-2) (Jerela., 2004; Andra et al., 2006).

Different compounds used for LPS inactivation
Lipid A neutralizing proteins like lactoferrin, lysozyme and collectins
Specific and cross-reactive anti-LPS antibodies
Intestinal alkaline phosphatase
Acyloxyacyl hydrolase (AOAH)
Lipopolyamines
Synthetic peptides like NK-lysin, cathelicidins
Endotoxoid based vaccines

The immune response against the microbes can either be non-specific, i.e., a non-specific resistance or an innate resistance, a natural immune response that the host exerts against any invading foreign body. The innate immune response is due to the general first line of defence of the host and includes barriers like the skin, mucus and lysozymes. The specific immune response, an acquired or adaptive immune response is specific against a particular foreign pathogen, such as bacteria, virus, or any toxin. Such immune response stores the memory of the infection. Such foreign antigens provoke the body to produce specific antibodies against them.

It has been documented that animals that survive exposure to endotoxin or Gram-negative bacteria often develop tolerance to subsequent challenges with LPS and other microbial agonists. This is by way of adaptive immunity, which is antigen-specific and requires the recognition of specific antigens via a process of antigen presentation that results in an immunological memory (Munford., 2010)

4.6. Genetic and Molecular Basis of Virulence

The new advances in various fields of biology and the advancements in molecular biology, genetics, and cell biology have completely transformed the way we look at the concept of the bacteria and host interactions and their basis of virulence. Areas such as protein expression profiling and techniques such as molecular crystallography have fundamentally altered our understanding of the host-pathogen interaction mechanism, providing a much more detailed picture of the overall scenario at the molecular and even atomic level.

However, the basic concept of pathogenicity as proposed by Smith depicts it as a multifactorial event that involves five basic steps (Weetman A 1992):

(i) Attachment
(ii) Entry
(iii) Multiplication
(iv) Interference with the defence system of the host
(v) Damage to the host

All of these events occur in a mutually dependent manner. The pathogen's specific factors mediate these steps and are therefore termed pathogenicity determinants. The expression of these determinants of pathogenicity usually occurs in a regulated fashion. The systems like environment sensing and quorum sensing allow the overall expression of these virulence factors.

4.6.1. Attachment

With the help of particular proteins identified in several bacterial species termed adhesins, the bacteria gets attached to the host cell membrane (Perrild et al. 1994, COOPER 1982, Williams et al. 1997 and Winsa et al. 1992). These adhesins proteins usually bind with their specific carbohydrate receptors on the host cell membrane. One of the most studied adhesions is *E.coli's* fimbrial adhesins. The fimbrial adhesins of *E.coli* are divided into two families:

1. The K88, K99, CFA/I and CFA/II adhesins: mediate attachment to the gut epithelium
2. Type 1. P and S fimbriae: mediate attachment to the urinary tract

The fimbriae are helical structures where major protein subunits support the minor protein subunits. The minor subunits of fimbriae help determine the receptor specificity of the attachment with urinary tract receptors, while in the case of gut epithelium, the major protein subunits help in determining the receptor specificity.

The genes of fimbriae had been successfully identified, cloned and sequenced. In some cases, even the individual proteins responsible for interaction with the host cell receptors have been identified for example, in the case of *S. fimbriae*, lysine and arginine residues at the 116 and 118 positions respectively were seen to be essential for its minor subunit (Glaser et al. 1997). Some host cell receptors have also been studied well at the molecular level. For example- the β-D-Gal groups are seen to be ligands for the K88 fimbriae.

Interestingly, a single receptor does not mediate this host-pathogen interaction, but many can be involved. For example, there are at least four types of receptors for *Streptococcus pneumonia*e with the respiratory tract, and its interaction depends explicitly on the organism's phenotype (LIPPE et al. 1987). Three interchangeable *Pneumococci* variants are distinguished by opaque, semitransparent, and transparent characteristics. The exact molecular basis for this type of phase variation is not known, but they differ in their ability to colonize the Nasopharynx (Weetman et al. 1994). The lung has two receptors at resting state GlcNAcβ1-3Gal and GlcNAcβ1-4Gal which are recognized by the transparent and opaque type of phenotypes. Cytokine-activated lung cells also express the platelet-activating factor receptor (PAF). The transparent variants can adhere to PAF, while the opaque ones cannot recognize PAF. Therefore, adhesion is a dynamic phenomenon wherein both the host and bacterial cell receptors differ based upon their site of isolation or the activation state of the cell.

4.6.2. Entry

The entry of the pathogen to the host cell can be direct with the invasion of the mucus membrane of the epithelial cells or passage between them followed by the invasion into the deeper tissues. Invasion is not a characteristic of all the pathogens like the *Vibrio cholerae* produces disease by the toxin production from the lumen of the intestine. *Shigella flexneri* invades the gut epithelium (Hashizume et al. 1991). Its entry is triggered by gene products – IpaB (invasion plasmid antigen B), IpaC (invasion plasmid antigen C) and IpaD (invasion plasmid antigen C). IpaB being haemolysin causes the release of *Shigella* from the vacuole into the cytoplasm from where the bacteria translocate towards host actin by polymerization that occurs due to the intracellular spread gene, i.e., icsA gene.

This movement leads to protrusions spanning from one cell to another. This protrusion the then ripped off to generate a vacuole inside the next cell. The double-layered membrane of this vacuole is then lysed with the product of the icsB gene. This causes the pathogen's release into the cytoplasm, and the cycle begins again.

4.6.3. Multiplication

Once into the host, the pathogen must multiply itself to cause its effect. The degree to which the pathogen could successfully multiply within the host determines its overall transmission potential to a new host. This will also determine the effect of its disease-causing potential of the pathogen as a rapid degree of multiplication corresponds to a higher degree of infection than a slow degree of multiplication. However, the exact molecular basis of multiplication events are still not known. Most of these pathogens utilize the nutrients from the host cell but the exact environmental and nutrient factors needed are not well understood. Most *in vivo* studies suggest that these molecular mechanisms involve overcoming iron restrictions. These pathogens need iron which they can obtain from different mechanisms like *S. pneumoniae* uses hemin or heme from haemoglobin (Cole et al. 1990) or indirect form from transferrin or lactoferrin in case of *Neisseria gonorrhoeae* (Goldsmith et al. 1987). In some indirect manner, as in case *E.coli*, they can utilize iron from the iron-binding proteins by producing siderophores (Thorburn et al. 1970) or in the case of *Mycobacterium tuberculosis* they use intracellular iron stores (Lepper et al. 1988).

4.7. Interference with the Defence System

For its survival within the host system, the pathogen must find its way to invade its immune system. For this purpose, the pathogen uses its various virulence factors of interference, including polysaccharide capsules, the protein toxins, and lipopolysaccharides. The process of phagocytosis and bacterial killing mediated via complement is usually interfered with the polysaccharide. The capsule mediated interference is extensively studied, for in the case of *pneumococcus*, it was seen that its non-capsular mutant was about one trillion times less virulent than the capsular ones (Watson et al. 1990). It was seen that sialyl groups on the polysaccharides of the capsule in *E.coli* and group B *streptococci* could prevent the complement activation pathway (Williams et al. 1997 and Jann et al. 1992). The host antibodies against capsular polysaccharides provide a protective immunity via opsonization of the pathogen by promoting phagocytosis. Pathogens in the capsulated form are the primary cause of most diseases as they could produce diverse capsular type ex. *Pneumococcus* or capsule making non-immunogenic as in the case of *E. coli* K1 and *Neisseria meningitides* group B. this non-immunogenicity can be due to the similarity between the molecular structure of polysaccharides subunits between pathogen and host cells.

The invading pathogen could also produce certain toxins that would interfere with the immune system response of the host, for example, the *Pneumococcus pneumolysin* toxin (Paton et al. 1993). These toxins have the potential to lyse all eukaryotic cells at high concentrations. At a lower concentration than the lytic concentrations, i.e., at sublytic levels, these toxins can show a diverse range of effects on the different cells and other soluble molecules part of the immune system. The pneumolysin, for example, inhibits the respiratory burst of phagocytes and prevents the random migration and chemotaxis of the cells (Paton et al. 1983). The toxins halt antibody production by the lymphocytes and the proliferation of B-cells induced by mitogens. In the *in vivo* studies, the toxins tend to induce a tremendous inflammatory response, undoubtedly maybe due to the production of inflammatory cytokines like interleukin 1-β and TNF-α (Houldsworth et al. 1994) and activate the classical complement pathway as well. The classical complement pathway activation is probably due to the characteristic property of the toxin to bind to the IgG antibody Fc portion. By the gene replacement studies, the role of toxins in the whole bacterium has been elucidated in *Pneumococcus* by constructing its variants expressing the altered version of toxin (Rubins et al. 1996 and Berry et al. 1995). Yet the molecular basis of the lipopolysaccharide activity studies are not as advanced as that of the protein virulence factors. It is now known that long O side chains of *E.coli* and *Salmonella* are needed for serum resistance.

4.8. Damage to the Host

The pathogen can directly damage the host cell by producing toxins or by inflammation and immunopathologic reactions. Some of the toxin effects at the molecular level have been thoroughly studied, as in cholera and tetanus. In the case of cholera toxin, it is an ADP ribosylating toxin wherein the enzymatic action of the toxin causes ADP ribosylation of a regulatory G protein of the adenylate cyclase complex enterocytes. This causes an increase in the levels of cAMP, which further alters the transport of ions across the epithelium and causes diarrhoea. Even the heat liable toxin of *E.coli* works similarly. Similar examples of the ADP ribosylating toxins include pertussis, diphtheria, and toxin A of the *Pseudomonas aeruginosa* but their overall effects vary according to the difference in their cellular targets the ADP ribosylation. Otherwise, the toxins can also be proteases like the tetanus toxin, a zinc protease that cleaves off synaptobrevin, a protein involved in releasing neurotransmitters. Other toxins are damaging membrane toxins, including the pore-forming proteins of *Staphylococcus aureus* alpha-toxin and thiol-activated toxins, including pneumolysin. Inappropriate or excessive production of the cytokines or complement pathway activation may lead to the stimulation of the inflammation. This may be due to the lipopolysaccharide production of Gram-negative bacterial cell walls that mediates the endotoxic shock. The molecular mechanism of this endotoxic shock is now beginning to be understood. This immunopathologic reaction may potentially even cause damage to the host (LIPPE et al. 1987 and Zychlinsky et al. 1995). The reactions cause antibody production that is cross-reactive to certain biomolecules, as seen in the case of infection with group A *streptococci* causing endocarditis. The molecular mechanism for such infection is partially known due to the M-protein of the *streptococci* that share epitopes with the antigens expressed in the heart tissue (Dale et al. 1982).

4.9. Virulence Factors Required to Infect the Host

In order to infect its specific host, the pathogen produces certain factors that lead to causing disease, these factors are collectively termed virulence factors. The virulence factors may include certain toxins, coat on pathogen surface which prevents their phagocytosis or certain surface receptors that could bind with host cells. Therefore, such virulence factors produced by the bacteria help the pathogen colonize into its host and hijack the host's defence mechanism.

These factors can be either secretory or membrane-associated or cytosolic (Sharma et al. 2016). It is not a mandate that the virulence factor produced should by produced by every pathogen strain. In most cases, the different bacterial strains have evolved mechanisms of different types of virulence factor production. For example, only some of the *E.coli* strains can produce enterotoxins which could cause diarrhoea (Sack 1975).

The infection is an altogether result of the misbalance between the host's resistance and the pathogenicity of the pathogen. However, the property of virulence doesn't solely depend upon the pathogen, but it's an overall interaction of the virulence factors with the host's defence responses. In a convenient viewpoint, the main aim of the pathogen is to survive within its host. Therefore, for most bacteria, their primary aim is to multiply within the host and therefore does not cause much harm to the host because simply if the host dies, the pathogen's life cycle would also end. Thus, the primary aim of the bacteria is to multiply. Bacteria being small in size, have various environmental favours for their survival and have a very high growth rate and substrate utilization capacity compared to the eukaryotic cells. Due to their high mutation rate and a short generation period, they tend to adapt better to survive in harsh environments. Therefore they usually cause mild symptoms to the host they infect, for example- *Rickettsia akari*, the causing agent of rickettsialpox only produces mild infection with a mild headache and fever (Peterson et al. 1996).

The different variety of pathogen could have a different range of virulence factors that they produce, which can be broadly discussed as:

4.9.1. Virulence Factors related to Adherence and Colonization:

There happens to occur a covering of mucosa over the host surface. In order to infect any organism, the pathogen has first to overcome this and adhere to the mucosal surface. The main problem here is that the mucosal lining is continuously flushed and replaced by a new lining released by the goblet cells. Similarly, the ciliated respiratory tract cells sweep the bacteria along with the mucus. The turnover rates of the epithelium lining present at these surfaces are rapids. Therefore, to successfully infect the host, the bacteria must tightly adhere to this surface and divide before the mucus is replaced with a new lining swiping the bacteria with it. Therefore, with the due course of evolution, the bacteria have developed mechanisms to overcome such situations in the form of certain organelles that aid their attachment such as the pili or fimbriae. Numerous colonization factors assist bacterial adherence. Some examples of bacteria with pili are *E.coli*, *V.cholera*, *S. pyogenes*, etc. (Peterson et al. 1996 and Van 1977).

4.9.2. Virulence Factors for Invasion:

The bacteria that infect the host by invading into their cells are usually either obligate intracellular types or can be primarily facultative intracellular types. The main factors responsible for invasion are unknown, yet there is evidence that numerous gene products are involved in the process. It was seen that the *Shigella* invasion factor, when conjugated into *E.coli* provided the property of invasion to them. Some more invasion factors have been recently recognized in few bacterial species yet many are still unknown (Sack 1975).

4.9.3. Capsules as Virulence Factors:

In order to survive in different environmental conditions, bacteria have evolved certain structural mechanisms too to survive well. Capsule formation is one such well-known mechanism. The bacteria that can encapsulate themselves, such as *Pneumococcus*, tend to overcome phagocytosis and intracellular killing by the host defence mechanisms. Some bacteria even have the property of serum resistance, like those bacteria that cause bacteremia. This property of serum resistance can be due to the capsular antigen composition and even the lipopolysaccharide structure.

There are even certain instances wherein specific bacterial pathogens tend to survive and divide within the phagocytic cells like the *Mycobacterium tuberculosis*, which is the causative agent of tuberculosis is an obligate intracellular parasite. It recognizes cell surface receptors like as phosphatidylinositol mannosidase (PIM), HSP70, 19 kDa lipoprotein, lipoarabinomannan (LAM) or certain pattern recognition receptors (PRRs) (Sharma et al. 2016). Its survival depends mainly upon the structure and composition of the cell surface (Peterson et al. 1996).

4.9.4. Endotoxins as Virulence Factor:

Endotoxin, defined first time in 1893 by Pfeiffer, refers to a toxic substance released after bacterial lysis. They are composed of lipopolysaccharides, the outer membrane of the Gram-positive bacteria. Endotoxins are one of the most extensively studied bacterial toxins and have different effects on the host besides giving an endotoxic shock (Peterson et al. 1996). Monophosphoryl lipid A being the most extensively studied endotoxin (Ribi et al. 1979).

Conclusion

The outermost layer of the bacterium's surface is where LPS is found when it comes into contact with different hosts. Any xenobiotic or compound can have a different effect on a different host. A host's age, family history, and immune system all play a role in the likelihood of infection. Obligatory and facultative intracellular bacteria are two types of bacteria that invade host cells and infect their hosts. Many gene products appear to be involved in invasion, but the exact role of each is still unclear. Toxins produced by gram-negative bacteria allow pathogens to bypass the body's innate immune system and infect organs far from their source. Disease outcomes can be influenced by which toxin is present. Understanding how these toxins affect host defence has improved greatly in recent years. More precise toxin targeting and better clinical outcomes could be the result of this.

References

Andrä J, Gutsmann T, Garidel P, Brandenburg K. Mechanisms of endotoxin neutralization by synthetic cationic compounds. *J Endotoxin Res.*, 2006; 12(5): 261–77.

Bates M, Akerlund J, Mittge E and Guillemin. Intestinal Alkaline Phosphatase Detoxifies Lipopolysaccharide and prevents inf;ammation in Zebrafish in response to the Gut Microbiota. *Cell Host & Microbe.*, 2007; 2(6): 371-382.

Berry A, Alexander J, Mitchell T, Andrew P, Hansman D, Paton J. Effect of defined point mutations in the pneumolysin gene on the virulence of Streptococcus pneumoniae. *Infection and Immunity.*, 1995; 63(5): 1969-1974.

Champion K, Chiu L, Ferbas J, Pepe M. Endotoxin Neutralization as a Biomonitor for Inflammatory Bowel Disease. *PLoS One.*, 2013; 8(6): e67736.

COOPER D. Agranulocytosis Associated with Antithyroid Drugs. *Annals of Internal Medicine.*, 1983; 98(1): 26.

Cole T, Hughes H. Sotos syndrome. *Journal of Medical Genetics.*, 1990; 27(9): 571-576.

Dale J, Beachey E. Protective antigenic determinant of streptococcal M protein shared with sarcolemmal membrane protein of human heart. *Journal of Experimental Medicine.*, 1982; 156(4): 1165-1176.

Glaser N, Styne D. Predictors of Early Remission of Hyperthyroidism in Children 384. *Pediatric Research.*, 1997; 41: 66-66.

Goldsmith M, Solish S, Voutilainen R, Miller W. LEYDIG CELL TUMOR PRESENTING AS CONGENITAL ADRENAL HYPERPLASIA (CAH). *Pediatric Research.*, 1987; 21(4): 247A-247A.

Hashizume K, Ichikawa K, Sakurai A, et al. Administration of thyroxine in treated Graves' disease. EVects on the levels on antibodies to thyroid stimulating hormone receptors and on the risk of recurrence of hyperthyroidism. *N Engl J Med*, 1991; 324: 947–53.

Houldsworth S, Andrew P, Mitchell T. Pneumolysin stimulates production of tumor necrosis factor alpha and interleukin-1 beta by human mononuclear phagocytes. *Infection and Immunity.*, 1994; 62(4): 1501-1503.

Jann K, Jann B. Capsules of Escherichia coli, expression and biological significance. *Canadian Journal of Microbiology.*, 1992; 38(7): 705-710.

Jerala R, Porro M. Endotoxin neutralizing peptides. *Curr Top Med Chem.*, 2004; 4(11): 1173–84.

Ji B, Jamet P, Perani EG, Bobin P, Grosset JH. Powerful bactericidal activities of clarythromycin and minocycline against *Mycobacterium leprae* in lepromatous leprosy. *J Infect Dis.*, 1993; 168: 188–90.

Lepper A, Wilks C. Intracellular iron storage and the pathogenesis of paratuberculosis. Comparative studies with other mycobacterial, parasitic or infectious conditions of veterinary importance. *Journal of Comparative Pathology.*, 1988; 98(1): 31-53.

LIPPE B, LANDAW E, KAPLAN S. Hyperthyroidism in Children Treated with Long Term Medical Therapy: Twenty-Five Percent Remission Every Two Years*. *The Journal of Clinical Endocrinology & Metabolism.*, 1987; 64(6): 1241-1245.

Mitchell T, Andrew P, Saunders F, Smith A, Boulnois G. Complement activation and antibody binding by pneumolysin via a region of the toxin homologous to a human acute-phase protein. *Molecular Microbiology.*, 1991; 5(8): 1883-1888.

Munford RS. Review detoxifying endotoxin: time, place and person. *J Endotoxin Res.*, 2005; 11(2): 69–84.

Munford RS. Murine responses to endotoxin: another dirty little secret? *J Infect Dis*, 2010, 201: 175–177.

Palsson-McDermott EM, O'Neill LA: Signal transduction by the lipopolysaccharide receptor, Toll-like receptor-4. *Immunology*, 2004, 113: 153–162.

Paton J, Andrew P, Boulnois G, Mitchell T. MOLECULAR ANALYSIS OF THE PATHOGENICITY OF STREPTOCOCCUS PNEUMONIAE: The Role of Pneumococcal Proteins. *Annual Review of Microbiology.*, 1993; 47(1): 89-115.

Paton J, Ferrante A. Inhibition of human polymorphonuclear leukocyte respiratory burst, bactericidal activity, and migration by pneumolysin. *Infection and Immunity.*, 1983; 41(3): 1212-1216.

Patrick M. Schlievert. Enhancement of Host Susceptibility to Lethal Endotoxin Shock by Staphylococcal Pyrogenic Exotoxin Type C. *Infection and Immunity.*, 1982; 36(1): 123-128.

Perrild H, Grüters-Kieslich A, Feldt-Rasmussen U, Grant D, Martino E, Kayser L et al. Diagnosis and treatment of thyrotoxicosis in childhood A European questionnaire study. *European Journal of Endocrinology.*, 1994; 131(5): 467-473.

Peterson JW. Bacterial Pathogenesis. In: Baron S, editor. Medical Microbiology. 4th edition. Galveston (TX): University of Texas Medical Branch at Galveston; 1996. Chapter 7. Available from: https://www.ncbi.nlm.nih.gov/books/NBK8526/.

Ribi E, Parker R, Strain SM, Mizuno Y, Nowotny A, Eschen KB, et al. Peptides as requirement for immuno therapy of the guinea-pig line-10 tumor with endotoxins. *Cancer Immunol Immunother.*, 1979; 7: 43 – 58.

Rubins J, Charboneau D, Fasching C, Berry A, Paton J, Alexander J et al. Distinct roles for pneumolysin's cytotoxic and complement activities in the pathogenesis of pneumococcal pneumonia. *American Journal of Respiratory and Critical Care Medicine*., 1996; 153(4): 1339-1346.

Sack R. HUMAN DIARRHEAL DISEASE CAUSED BY ENTEROTOXIGENIC ESCHERICHIA COLI. *Annual Review of Microbiology*., 1975; 29(1): 333-354.

Sharma A, Dhasmana N, Dubey N, Kumar N, Gangwal A, Gupta M et al. Bacterial Virulence Factors: Secreted for Survival. *Indian Journal of Microbiology*., 2016; 57(1): 1-10.

Thorburn M. Exomphalos-Macroglossia-Gigantism Syndrome in Jamaican Infants. *Archives of Pediatrics & Adolescent Medicine*., 1970; 119(4): 316.

Weetman A. How antithyroid drugs work in Graves' disease. *Clinical Endocrinology*., 1992; 37(4): 317-318.

Van Oss C. The Pathogenesis of Infectious Disease C.A. Mims. *Immunological Communications*., 1977; 6(4): 449-450.

Williams K, Nayak S, Becker D, Reyes J, Burmeister L. Fifty Years of Experience with Propylthiouracil-Associated Hepatotoxicity: What Have We Learned?1. *The Journal of Clinical Endocrinology & Metabolism*., 1997; 82(6): 1727-1733.

Winsa B, Dahlberg P, Jansson R, Ågren H, Karlsson F. Factors influencing the outcome of thyrostatic drug therapy in Graves' disease. *Acta Endocrinologica*., 1990; 122(6): 722-728.

Weetman AP 1, A P Pickerill, P Watson, V K Chatterjee, O M Edwards. Treatment of Graves' disease with the block-replace regimen of antithyroid drugs: the effect of treatment duration and immunogenetic susceptibility on relapse. *QJM: An International Journal of Medicine*., 1994 Jun; 87(6): 337-41.

Watson D, Musher D. Interruption of capsule production in Streptococcus pneumonia serotype 3 by insertion of transposon Tn916. *Infection and Immunity*., 1990; 58(9): 3135-3138.

Zychlinsky A. Virulence mechanisms of bacterial pathogens edited by J.A. Roth, C.A. Bolin, K.A. Brogden, F.C. Minion and M.J. Wannemuehler ASM Press, 1995. £52.50 hbk (xvii + 366 pages) ISBN 1 55581 085 3. *Trends in Microbiology*., 1996; 4(11): 459-460.

Chapter 5

Endotoxins as Vaccine Adjuvants

Sheza Farooq[1], Shikha Chaudhary[1], R. S. Sethi[2] and Arif Pandit[3,*]

[1]PhD Scholar, Department of Animal Biotechnology, College of Animal Biotechnology, Guru Angad Dev Veterinary and Animal Sciences University, Ludhiana, Punjab, India
[2]Professor-cum-Head, Department of Animal Biotechnology, College of Animal Biotechnology, Guru Angad Dev Veterinary and Animal Sciences University, Ludhiana, Punjab, India
[3]Assistant Director Research, Directorate of Research, Sher e Kashmir University of Agricultural Sciences and Technology of Kashmir, Kashmir, India

Abstract

Endotoxin lipopolysaccharides (LPS) are the major components of the exterior cell wall of Gram-negative bacteria. LPS typically consists of three parts, a polysaccharide O-antigen, a core-oligosaccharide and a hydrophobic Lipid-A. The Lipid-A component of LPS plays a significant role in facilitating the endotoxin's major bioactivity. It strongly activates the innate immune response and is a major determinant in the adaptive immune response. LPS or its derived molecules, if added to vaccines or the use of bacterial derived vaccines that naturally contain LPS, can behave as an adjuvant to elicit a more robust immune response. The structure of LPS can be extensively modified to produce better immune response and decrease toxicity. As these endotoxins are thermolabile, they can be inactivated by heat or chemical treatment, producing a toxoid that can be further deactivated and therefore used to produce vaccines adjuvants. In the seventies, a modified LPS named Monophosphoryl Lipid-A (MPLA) was developed by chemically detoxifying LPS from *Salmonella minnesota*. Since it is a Toll-like receptor-4 agonist, it is a potential candidate as a vaccine adjuvant. Further, it has reduced toxicity

[*] Corresponding Author's E-mail: arif.pandit@skuastkashmir.ac.in.

In: Endotoxins and their Importance
Editors: Arif Pandit and R. S. Sethi
ISBN: 978-1-68507-839-3

as in vitro data shows the release of pro-inflammatory cytokines by innate immune cells while retaining the adjuvant effect. The decreased toxicity is due to reduced signalling through MyD88 while retaining signalling through the TRIF pathway. However, the main drawback of MPLA is the possible activation/enhancement of TLR4- related autoimmune diseases and its elevated production and subsequent purification cost. A solution to this problem could be the generation of synthetic Lipid-A analogues of MPLA such as RC-529. Another way of detoxifying LPS is a genetic modification which can be achieved by inactivating or modifying genes involved in Lipid-A biosynthesis. Some bacteria, such *Bacteroides thetaiotaomicron* and *Prevotella intermedia,* naturally express an MPLA-like Lipid-A structure. An efficient vaccine can also be made by combining LPS with OMP, e.g., LPS-J5/OMP vaccine. To make this vaccine, the OMP of Neisseria meningitidis group B is non-covalently complexed with detoxified LPS. It eventually promotes Gram-negative bacterial clearance in vivo. The LPS-based structures are valuable adjuvants in vaccine manufacture and development.

Keywords: LPS, lipid-A, endotoxin, adjuvant

Introduction

Endotoxin is a lipopolysaccharide present in the outer membrane of the Gram-negative cell wall and constitutes 35–45% of the outer membrane. LPS imparts great structural integrity to bacterial cells. It contains Lipid-A portion consisting of fatty acids and disaccharide phosphates, core polysaccharides, and the O-antigen. Lipid-A, a fraction of LPS has endotoxin activity identified by mammalian immune cells through the pattern recognition receptor complex toll like receptor 4 (TLR4)/myeloid differentiation protein 2 (MD-2). This causes activation of immune cells and inflammatory cytokines to be released. TLR4 stimulation causes intracellular protein complex formation, resulting in the activation of intracellular signalling pathways (Kawai et al., 2006 and Miggin et al., 2006). These reactions cause the biosynthesis and secretion of a variety of pro-inflammatory cytokines (IL-1, IL-8, IL-12, TNF, and IFN) as well as the production of co-stimulatory molecules (Alexander et al., 2011), which in turn activate humoral and cellular responses such as complement activation (Morrison et al., 1997 and Cooper et al., 1978). As a result, this response is useful in infection control in the local area. There are ten known human TLRs, each of which identifies different types of PAMPs like bacterial lipoproteins, bacterial DNA, viral RNA, or bacterial flagella. TLR1, 2, 4, 5, and 6 are found on the cell surface, whereas TLR3, 7, 8, and 9 are found in endosomal compartments, albeit TLR4 can migrate to and signal from the endosomal compartment after ligand recognition. (Guo et al., 1998). TLRs produce homo- or heterodimers in response to ligand recognition, triggering a signalling cascade. Adaptor molecules such as myeloid differentiation primary response gene 88 (MyD88), Toll/IL-1 receptor (TIR)-domain-containing adapter-inducing IFN-b (TRIF), TIR-containing adaptor protein, and TRIF-related adaptor molecule are typically recruited first in this signalling cascade. (Guo et al., 1998 and Miller et al., 2005). Except for TLR3, all TLRs signal through MyD88. Signalling through MyD88 occurs with the inclusion of TIR-containing adaptor protein in the case of TLR1, TLR2, and TLR6. TLR3 only employs TRIF to communicate, but TLR4 uses a TIR-containing adaptor protein to bind MyD88 on the cell surface and a TRIF-related adaptor molecule to bind TRIF in the endosome. As a result, TLR signalling can be classified as either MyD88 dependent or independent (TRIF dependent). The IL-1 receptor-associated kinase family members are recruited by MyD88-dependent signalling, which activates the kinases TNF-associated factor 6 (TRAF 6), TGF-b–activated kinase 1 and the IkB kinase (IKK) complex, resulting in the translocation of NF-kB to the nucleus and the transcription of pro-inflammatory cytokines. LPS is active pathogen-associated molecular patterns (PAMPs) and an inducer of innate immunity. Because of the high safety demands, increased knowledge, and concerns about side effects, the implementation of new techniques and vaccine development has shifted away from inactivated whole organisms towards the highly purified recombinant proteins and single antigens. Until recently, the only way to induce prophylactic immunity to human infections was to use infectious-attenuated or inactivated entire viral particles or bacteria. In the broader context of public health, these vaccinations have made and continue to make remarkable contributions to eradicating harmful or life-threatening diseases. However, the reliance on vaccines alone has become restricted, necessitating adjuvants' inclusion. Adjuvants are nothing but molecules that provide an antigen. According to available information, adjuvants trigger immunological responses via multiple mechanisms like their depot effect (sustained release of the antigen), by enhancing the uptake of antigen, or by activation and maturation of MHCs. Adjuvants are also responsible for the upregulation of cytokines and chemokines and the recruitment of cells at the injection site. This benefit was initially used to enhance the efficacy of inactivated whole pathogen vaccines but is now considered a virtual necessity in the context of subunit vaccination. Alum has been used in vaccinations for almost 70 years and has been utilised in hundreds of millions of doses (Garcon et al., 2007; Brewer, 2006, and Glenny et al., 1926). This extensive experience demonstrates that Alum is safe but with few adverse effects. However, it primarily stimulates the synthesis of IgG4 and IgE isotype antibodies, better adapted for responses against external pathogens and parasites rather than destroying or phagocytosing pathogen-infected host cells. Alum causes a Gr-1+ subset of leukocytes to produce substantial levels of IL-4, a conventional Th2 cytokine, and Gr-1+ eosinophils to be recruited within 6 hours of alum and antigen injection in experimental mice. The Th2 bias associated with the use of Alum could be explained by the early recruitment of IL-4 generating eosinophils. In adjuvant design, two factors have combined to produce the necessity to go beyond Alum. First, for infectious illnesses that are recognised as global health problems but for which vaccine research has been ineffective or only partially effective, such as tuberculosis, HIV/AIDS, and malaria, a requirement for Th1-type immunity is widely acknowledged (Garcon et al., 2007; Golkar et al., 2007; Pulendran and Ahmed, 2006). LPS is an effective inducer of innate immunity and an effective booster of the immunological response to protein antigens. LPS can also produce a bias in the T cell response toward Th1 (Mcaleer et al., 2008). However, LPS's utility as an immune adjuvant is severely restricted due to its endotoxic action, resulting in inadequate reactogenicity. Following that, efforts were made to attenuate LPS and separate endotoxic and immunostimulatory effects, either through direct chemical treatment of pure LPS or through genetic regulation of LPS synthesis. However, LPS has high toxicity that limits its therapeutic application in humans. LPS based compounds with lower toxicity but retained adjuvant characteristics were created by altering its chemical structure. The most successful LPS-based adjuvant, Monophosphoryl Lipid-A (MPLA), has been widely employed in vaccine trials and preclinical investigations. MPLA can be the first of a new family of TLR-stimulatory vaccination adjuvants to be commonly used in humans (Casella and Mitchell, 2008). Developing novel detoxified LPS species with better adjuvant properties than Alum, and lower toxicity has opened new immunisation possibilities.

This book chapter introduces LPS structure, function, toxicity, and immune system activation. Next, Lipid-A biosynthesis, its structure and roles as adjuvant have been discussed, followed by the structure of endotoxin receptor. A final overview section summarises the most relevant points.

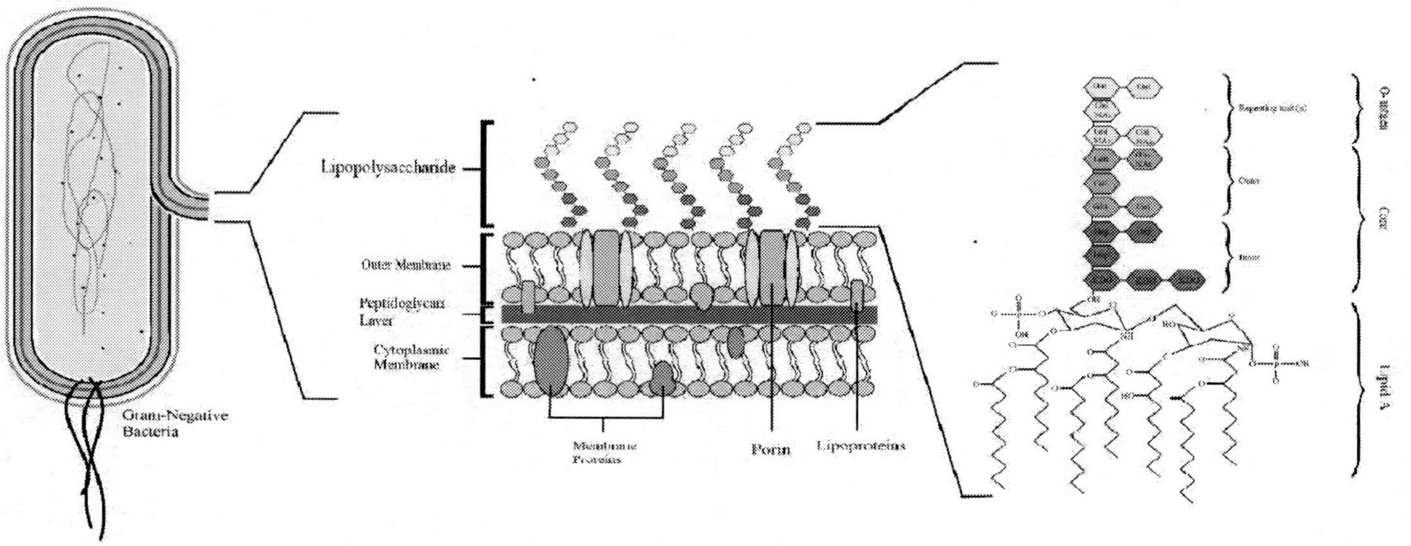

Figure 5.1. The LPS structure of Gram-negative bacteria (*Escherichia coli*) Lipid-A.

5.1. LPS Structure and Function

Lipopolysaccharide (LPS), often known as endotoxin, is Gram-negative bacteria's primary cell wall element. Its composition varies depending on the bacteria type from which it is produced. Gram-negative bacteria's outer membrane (OM) is a bilayer of glycerophospholipids on the inner surface and primarily LPS on the outer surface. It was proposed in the 1960s that LPS had a typical general architecture: core-oligosaccharide, Lipid-A, and O-specific side chain (Caroff et al., 2002; Caroff et al., 2003). A more elaborate structure of LPS was discovered later using mass spectrometry and high-field nuclear magnetic resonance spectroscopy. The hydrophobic part of LPS that is anchored in the OM is called Lipid-A. Its basic structure is a glucosamine disaccharide with four to six acyl chains and two phosphate groups. β-1, 6-linked d-glucosamine disaccharide is coupled to a variable number of ester- and amide-linked molecules in the Lipid-A domain (Figure 1).

Its structure is highly conserved, but the quantity and fatty acid side chains length, terminal phosphate residues, and related modification may differ amongst bacteria. If present, the O-antigen is highly variable out of the three domains, consisting of 50 repeating oligosaccharide units of 2–8 monosaccharide subunits. Certain gram-negative bacteria that colonise mucosal surfaces, such as Neisseria meningitides, lack this O-chain, and the resulting molecule is referred to as lipo-oligosaccharide rather than LPS (Jennings et al., 1980). A potential outcome of gram-negative infection is septic shock, of which many symptoms can be attributed to LPS, more precisely, the Lipid-A of LPS. In addition, certain modifying enzymes can alter the LPS composition, increasing the LPS heterogeneity (Guo et al., 1998; Trent et al., 2001). Toll-like receptors (TLR)s belong to the receptor's family that recognise a broad diversity of specific but conserved structures of pathogen microorganisms. Immediately after stimulation, TLRs activate immune defence mechanisms. Toll-like receptor 4 (TLR4), in conjugation with the glycoprotein MD-2, constitutes the LPS receptor (Plasson et al., 2004; Trinchieri et al., 2007). The Lipid-A region mediates TLR4 stimulation. In this regard, variations in its structure, particularly the length and amount of acyl acid chains and charge, are critical (Miller et al., 2005). Hexa-acylated E. coli Lipid-A (canonical LPS structure, shown in Figure 1), which comprises fatty acids with 12–14 carbons and two phosphate residues, is the most potent activator of human TLR4 (hTLR4) (Golenbock al., 1991; Kotani et al., 1985) The tetra-acylated Lipid IVa with 18–16 carbon fatty acids and a phosphate residue, an intermediary in the production route of Lipid-A, does not induce hTLR4 (canonical hTLR4 antagonist) (Saitoh et al., 2004).

5.2. Lipid-a Biosynthesis

The hydrophobic component of LPS (Lipid-A), is responsible for activating most of its innate immunity. TLR4/MD-2 detects Lipid-A, and TLR4s dimerisation culminates in signalling via adaptor molecules MyD88 and TRIF. Finally, signalling via adaptor molecules and NF-kB translocation triggers the production of the inflammatory cytokines like IL-6 and TNF-a. The pathway of Lipid-A biosynthesis has been found in *Arabidopsis thaliana*, and it is quite like the *E. coli* Lipid-A biosynthesis pathway (Li et al., 2011). UDP-N-acetylglucosamine (GlcNAc) is the initial precursor molecule required for Lipid-A synthesis, and LpxA catalyses the first step in the route, fatty acylation of UDP-GlcNAc. In the first step of Lipid-A biosynthesis, LpxC deacylates UDP-3-O-(acyl)-GlcNAc. LpxD adds a second R-3-hydroxymyristoyl chain to generate UDP-2,3-diacylGlcN in the next step. Pyrophosphate linkage of UDP-2,3-diacyl-GlcN is cleaved by LpxH to generate 2,3-diacyl-GlcN-1-phosphate (lipid X) and the 1–6–linked disaccharide found in all lipids. LpxB then generates a molecule by condensing lipid X with UDP-2,3-diacyl-GlcN and releasing UDP. The final four stages of the Kdo2-Lipid-A synthesis pathway are catalysed by LpxK, KdtA, LpxL, and LpxM. LpxK phosphorylates disaccharide 1-phosphate at position 4' to produce Lipid IVa, the simplest LPS structure required for *E. coli* survival. KdtA usually adds at least two Kdo residues, while it may only add one in some species, including *Haemophilus influenzae* and *Bordetella pertussis* (Li et al., 1978; Zanze et al., 1987). LpxL and LpxM, respectively, add secondary lauroyl and myristoyl residues to the primary 3-OH fatty acid chains at 2' and 3' positions of Lipid-A in the last steps of Kdo2-Lipid-A biosynthesis (Raetz et al., 2002; Raetz et al., 2007). In the lipid synthesis process, there are a few biosynthetic steps. Bacterial strains with a changed Lipid-A structure emerge when a pathway is shut off. Inactivation of the LpxL and LpxM homologs, for example, lead to the loss of one or both secondary acyl chains in most bacteria (Vvan-der-ley et al., 2001). However, because most of the early processes are necessary for bacterial survival, knocking off those genes is not an option. *Moraxella catarrhalis* and *Acinetobacter baumannii* are two other gram-negative bacterial species that have been shown to persist in the absence of LPS (Moffatt et al., 2010; Peng et al., 2005).

Naturally occurring LPS-deficient mutants have only been described for *A. baumannii* and *N. meningitidis*, where the bacteria spontaneously turned off LPS production by mutation of their lpxD or lpxH genes, respectively (Piet et al., 2014).

5.3. Immunobiology of LPS

The host immune system has developed to recognise conserved bacterial molecular patterns as indicators of bacterial infection, including bacterial cell wall/outer membrane components, LPS in Gram-negative bacteria and lipoteichoic acid (LTA) in Gram-positive bacteria (Ulevitch et al., 1999). These components are detected as pathogen-associated molecular patterns (PAMPs) by pattern recognition receptors (PRRs), which include Toll-like receptors, when bacteria release them as a result of cell division, death, or, in particular, antibiotic treatment against bacterial infection (Hopkin et al., 1978 and Ginsburg et al., 2002). These receptors are expressed on innate immune cells, primarily mononuclear phagocytes (monocytes and macrophages), resulting in increased phagocytic activity, secretion of pro-inflammatory cytokines like tumour necrosis factor-alpha (TNF-α), interleukin 6 (IL-6), and induction of other pro-inflammatory proteins like inducible NO synthase (iNOS) (Rietschel et al., 1996; Dobrovolskaia 2002; Poltoraket al., 1998; Schwandner et al., 1999 and Palsson-McDermot et al., 2004).

PAMPs activate the immune system, allowing the organism to mobilise all of its resources to combat the invading pathogen at the right time and in the correct place. However, uncontrolled and sustained activation induces a systemic and persistent immunological response in some situations. High blood levels of pro-inflammatory cytokines, particularly TNF-α, characterise this systemic reaction (Woltmann et al. 1998 and Cohen et al., 2002). TNF-α is the primary mediator responsible for the pathogenic effects of endotoxic shock. Endothelial injury, loss of vascular tone, coagulopathy and multiple-system organ failure are among them, with the last typically culminating in death. (Hardaway et al., 2000).

5.4. Lipid-A Analogous Structures and Their Role as Adjuvants

Vaccines that are based on attenuated or inactivated whole pathogens have a wide range of TLR target structures, resulting in strong, robust, and long-lasting protection. In contrast, vaccines composed of one or a few purified components (subunit vaccines) are safe but have low immunogenicity, necessitating the addition of immune stimulants (adjuvants). The most used adjuvant is Alum, which refers to numerous aluminium salts. It induces a Th2-type antibody response, which has been efficacious in the vaccination (Brewer et al., 1999). On the other hand, Alum does not appear to stimulate Th1-type antibody responses (Dubanski et al., 2010). However, it predominantly promotes a Th2-type antibody response (Brewer et al., 2006 and McKee et al., 2007) consisting of IgG4 and IgE isotypes, which are best suited for responses against extracellular pathogens and parasites rather than killing or phagocytosing pathogen-infected host cells. The specific bases for Alum's Th2-bias are beginning to be elucidated. Alum stimulates a Gr-1+ subpopulation of leukocytes in experimental animals to produce enormous amounts of IL-4 (Jordan et al., 2004), a canonical Th2 cytokine and Gr-1+ eosinophils recruited within 6 hours of alum and antigen injection (Kool et al., 2008). This early recruitment of IL-4 producing eosinophils could explain the Th2 bias associated with use of Alum. To generate maximum immune protection against infections or illnesses like cancer or allergies, adjuvants favour Thl or more balanced Th1/Th2 responses are necessary. Besides, Alum has a weak effect on mucosal immunity. Many pathogens attack mucosa tissues, providing an ecological niche for commensal and opportunistic microbes like Neisseria meningitides.

Consequently, mucosal immunity is the gold standard for preventing pathogen colonisation and conferring herd immunity against specific diseases. Vaccine adjuvants that target mucosal immunisation must encourage various biological and complicated processes, including Th17 cell growth, APC proliferation, and IgA synthesis (Chen et al., 2010; Lawson et al., 2011). Three new adjuvants were licenced as alternatives to Alum: MF59, AS03, and RC-529. MF59 is a low-oil-content oil-in-water emulsion that is incorporated in an authorised influenza vaccine (O'hagan, 2007). MF59 elicits a more balanced Th1/Th2 response than Alum, although it is only partially effective, necessitating the use of Th1 enhancers. Because of its remarkable ability to generate Th1-type responses against co-administered antigens, LPS has gained much interest due to its adjuvant properties. IL-1 plays a vital role in the development of this response. Because TLR4 receptors are widely positioned on mucosal surfaces, TLR4 agonists should stimulate immune responses at both local and distant mucosal sites. Chemical treatments to change the LPS composition, notably Lipid-A, and chemical production of Lipid-A analogues, are two examples of strategies utilised in recent decades to reduce its high toxicity without impacting this core potential.

MPL is now utilised as an adjuvant in human vaccine preparations in Europe (human papillomavirus (Cervarix) and pollen allergy (Pollinex Quattro) (Drachenberg et al., 2001; Drachenberg et al., 2003) and Australia (hepatitis B virus) (Fendrix), and it has been extensively employed in human vaccination studies for numerous infectious diseases such as malaria (Aide t al., 2011), TB (Von et al., 2009; Polhemus et al., 2009) or tumour growth (Brichard et al., 2008). Because MPL is highly hydrophobic and leads to the production of aggregates in aqueous solution, which can have a significant impact on TLR4 activation, it is frequently combined with alum or other delivery vehicles (Garcon et al., 2007). When combined with other factors (accompanying antigen or delivery route), these combinations can change the adjuvant's activity. MPL, for example, boosts antibody formation in aqueous formulations while stimulating T-cell responses in oil in water emulsions. Nevertheless, MPL is a potent activator of cytotoxic T lymphocyte proliferation when paired with other delivery systems. The biological properties of delivery systems can also be altered. Liposomes are spherical vesicles made up of phospholipid bilayers widely used to deliver antigens in their native form. The inclusion of MPL into liposomes lowered the residual toxicity of MPL while preserving its adjuvant potential (Fries et al., 1992). This effect was also noticed with other detoxified LPS species (Aranas et al., 2010). As a result, liposomal MPL formulations have been extensively explored in human trials for various indications such as malaria, pneumococcal disease (Vernacchi et al., 2002) or genital herpes type 2 (Olson et al., 2009), as well as in experimental animals for Streptococcus pyogenic infections (Hall et al., 2004) or *E. coli* toxin neutralisation (Tana et al., 2003). Finally, MPL's efficacy in producing mucosal immunity following mucosal (Baldridge et al., 2000) and intramuscular (Kidon et al., 2011; Labadie et al., 2011) delivery is noteworthy; MPL formulations for mucosal vaccines have been intensively investigated for the treatment of several diseases, including genital herpes (Morello et al., 2011) and mucosal leishmaniasis (Llanos-cuentas et al., 2010).

5.5. Structure of the Endotoxin Receptor

LPS and MPLA require TLR4 receptors for adjuvant action (Evans et al., 2003, Qureshi et al., 1999; Muta et al., 2001; Takeuchi et al., 1999 and Hirschfeld et al., 2000). In most cases, LPS-binding protein (LBP) in the serum identifies LPS monomers. CD14 catalyses the LPS transfer from LBP to MD2 (the receptor system's LPS-binding component), and MD2 subsequently prompts TLR4 to signal. MD2 binds the acyl chains of its Lipid-A ligands with an elongated pocket shape whose inner face is lined with hydrophobic residues. Kim et al. (2007) postulated that LPS interaction produced a conformational shift in MD2, causing each to attach to two TLR4 molecules simultaneously. It was also found that heterotetrametric structures were formed when LPS was added to MD2/TLR4 heterodimers. Thus, it was possible to create a heterotetramer comprised of two LPS-loaded MD2 molecules linked to two TLR4 molecules. It is likely that the monophosphate structure of MPLA, or the higher-order aggregates described by Triantafilou et al., causes this heterotetramerization.

5.6. Live Vaccines Based on Lipid-a Mutants

The live vaccines have two primary advantages: self-adjuvating capability and the expression of the entire pathogen genome, which provides a diverse array of antigens. The strain must be attenuated to make live pathogen safe as a vaccine. Repeated passages on a cell line or animal embryos are the traditional methods for attenuating a strain. Live attenuated vaccines for *F. tularensis* and *Y. pestis*, the causative agents of tularemia and plague, respectively, have been created in this manner. Although it provides moderate protection against the virulent strain, the tularemia vaccine was not approved due to a lack of understanding of attenuation, and colony type instability (Marohn et al., 2013). An *F. novicida* mutant missing the Lipid-A 4'-phosphatase LpxF has been studied as an alternate method of attenuation (Bang et al., 2007). Since LpxF-mediated dephosphorylation of Lipid-A impairs the activation of the innate immune system, lpxF mutants may be effective for tularemia immunisation.

Reactogenicity is caused due to bacterial lysis within the host. LPS is a potent inducer of this reactogenicity. Therefore, lowering its TLR4-activating capacity can also be applied to live attenuated vaccines. The Lipid-A modifying enzyme LpxE from *F. tularensis* is expressed in *Salmonella* and removes the 1-phosphate group of Lipid-A explicitly. When combined with

PagP deletion, this change results in 4'-monophosphoryl hexa-acylated Lipid-A (Kong et al., 2011). Mutation of lpxM, which encodes an enzyme for adding a secondary myristic acid to Lipid-A, has also been studied for lowering reactogenicity. A lpxM mutation in a tumour-targeting *Salmonella* strain reduced TNF-a production and pathogenicity while keeping the parent strain's tumour-targeting capacity (Low et al., 1999).

The stimulation of MyD88-dependent and -independent (TRIF-dependent) cytokines was studied in LPS preparations from a few bacterial species. In macrophages, *N. meningitidis* LPS is a potent inducer of both pathways, *E. coli* O55:B5 and MyD88-dependent pathway, and Salmonella LPS typically triggers the TRIF pathway (Zughaier et al., 2005).

Chemical detoxification *Salmonella minnesota* LPS was first attempted in the 1970s. It typically contains up to three phosphates, seven acyl chains, and a polysaccharide chain of variable length. A combination of acylated di-glucosamines with just one phosphoryl group at the 4'position, a main species of six acyl chains, and no polysaccharide, occurred from sequential acid and alkaline hydrolyses of this LPS. MPLA is the term given to this modified LPS. MPLA is believed to reduce toxicity *in vitro* by inhibiting the production of pro-inflammatory cytokines by innate immune cells while maintaining a distinct adjuvant activity. MPLA had no adverse effects in animal models and was regarded as a safe adjuvant based on clinical investigations (Ulrich et al., 1995; Chilton et al., 2013).

Certain types of bacteria typically produce a comparatively less toxic LPS, making them possible candidates for whole-cell vaccines with intrinsic adjuvant activity or as an origin of isolated LPS for adjuvant application. *For example, Bacteroides thetaiotaomicron and Prevotella intermedia* naturally express a monophosphorylated and penta-acylated Lipid-A structure identical to MPLA. Genetic editing is another technique to detoxify LPS. This can be accomplished by inactivating or modifying genes involved in lipid metabolism. In N. meningitidis, the proteins LpxL1 and LpxL2 add a secondary lauroyl residue, and their inactivation causes viable strains to produce a penta-acylated Lipid-A structure (Van-der-ley et al., 2001). A biosynthesis, in particular genes involved in the addition of secondary acyl chains, is not required for bacterial viability, contrary to previous steps in the biosynthesis process.

Conclusion

Although LPS has been identified as an immune stimulant with potential adjuvant applications, its clinical use has been severely limited due to many unfavourable side effects. On the other hand, the discovery that MPL was safe and retained the desired adjuvant properties of LPS opened up new avenues for treating pathogen illnesses. Unlike previous adjuvants, LPS-based adjuvants provide novel benefits by stimulating cytotoxic T cells and enhancing Th1-type responses. This activity is required to protect against a variety of infections and to develop prophylactic medicines for diseases such as cancer and allergies. Indeed, MPL's success in existing vaccines where traditional adjuvants failed to provide protection backs this up.

Furthermore, LPS-based adjuvants' high adjuvant capacity has significant and obvious benefits in mucosal protection, lower booster doses, functional vaccination in the elderly, and the development of polyvalent vaccine formulations. MPL has several drawbacks, including higher production costs and the potential for TLR4-related autoimmune disorders to be activated or enhanced. The production costs of lipid-synthesised analogues with similar biological activity, such as MPL, have been significantly reduced. The data gathered so far from human immunisation provides additional safety proof regarding TLR4 autoimmune disease activation. Finally, LPS-based adjuvants improve current vaccination regimens and open up new avenues for solving problems.

References

[1] Aide P, et al. Four-year immunogenicity of the RTS, S/AS02(A) malaria vaccine in Mozambican children during a phase IIb trial. *Vaccine.* 2011;29:6059–6067.

[2] Alexander C, Rietschel ET: Bacterial lipopolysaccharides and innate immunity. *J. Endotoxin Res.* 2001; 7, 167–202.

[3] Arenas J, et al. Coincorporation of LpxL1 and PagL mutant lipopolysaccharides into liposomes with Neisseria meningitidis opacity protein: Influence on endotoxic and adjuvant activity. *Clin. Vaccine Immunol.* 2010;17 487–495.

[4] Baldridge JR, Yorgensen Y, Ward JR, Ulrich, JT. Monophosphoryl Lipid-A enhances mucosal and systemic immunity to vaccine antigens following intranasal administration. *Vaccine.* 2000;18:2416–2425.

[5] Brewer JM. (How) do aluminium adjuvants work? *Immunol Lett* 2006; 102:10–15.

[6] Brewer JM. et al. Aluminium hydroxide adjuvant initiates strong antigen-specifi c Th2 responses in the absence of IL-4- or IL-13-mediated signaling. *J. Immunol.* 1999;163: 6448–6454.

[7] Brichard VG, Lejeune, D. Cancer immunotherapy targeting tumour-specifi c antigens: towards a new therapy for minimal residual disease. *Expert Opin. Biol. Ther.* 2008; 8: 951–968.

[8] Caroff M, Karibian D, Cavaillon JM, et al. Structural and functional analyses of bacterial lipopolysaccharides. *Microbes Infect* 2002;4(9):915-26.

[9] Caroff M, Karibian D. Structure of bacterial lipopolysaccharides. *Carbohydr Res* 2003; 338(23):2431-47.

[10] Casella CR and Mitchell TC. Putting endotoxin to work for us: monophosphoryl Lipid-A as a safe and effective vaccine adjuvant. *Cell Mol Life Sci.* 2008;3231-40.

[11] Chen K, Cerutti A. Vaccination strategies to promote mucosal antibody responses. *Immunity.* 2010;33:479–491.

[12] Chilton PM, Hadel DM, To TT, et al. Adjuvant activity of naturally occurring monophosphoryl lipopolysaccharide preparations from mucosa-associated bacteria. *Infect Immun* 2013;81(9):3317-25.

[13] Cohen J, The immunopathogenesis of sepsis, *Nature.* 2002; 420:885–891.

[14] Cooper NR, Morrison DC: Binding and activation of the first component of human complement by the lipid A region of lipopolysaccharides. *J. Immunol.* 1978; 120: 1862–1868.

[15] Dobrovolskaia MA, Vogel SN, Toll receptors, CD14, and macrophage activation and deactivation by LPS, *Microbes Infect.* 2002;4: 903–914.

[16] Drachenberg KJ, Heinzkill M, Urban E, Woroniecki SR. Efficacy and tolerability of shortterm specific immunotherapy with pollen allergoids adjuvanted by monophosphoryl Lipid-A (MPL) for children and adolescents. *Allergol. Immunopathol.* (Madr.). 2003;31: 270–277.

[17] Drachenberg KJ, Wheeler AW, Stuebner P, Horak F. A well-tolerated grass pollen-specifi c allergy vaccine containing a novel adjuvant, monophosphoryl Lipid-A, reduces allergic symptoms after only four preseasonal injections. *Allergy.* 2001;498–505.

[18] Dubensky Jr TW, Reed SG. Adjuvants for cancer vaccines. *Semin. Immunol.* 2010; 22: 155–161.

[19] Evans JT, Cluff CW, Johnson DA, Lacy MJ, Persing DH, and Baldridge JR. Enhancement of antigenspecific immunity via the TLR4 ligands MPL adjuvant and Ribi.529. *Expert. Rev. Vaccines.* 2003;2:219 – 229.

[20] Franchi L, Park, JH, Shaw MH, Marina-Garcia N, Chen G, Kim YG, and Nunez G. Intracellular NOD-like receptors in innate immunity, infection and disease. *Cell. Microbiol.* 2008;10:1 – 8.

[21] Fries LF, et al. *Liposomal malaria vaccine in humans: a safe and potent adjuvant strategy.* Proc. Natl. Acad. Sci. U. S. A. 1992;89:358–362.

[22] Garcon N, Chomez P, and Van MM. GlaxoSmithKline Adjuvant Systems in vaccines: concepts, achievements and perspectives. *Expert. Rev. Vaccines.* 2007; 6:723–739.

[23] Ginsburg, Role of lipoteichoic acid in infection and inflammation, Lancet, *Infect. Dis.* 2.2002; 171–179.

[24] Ginsburg, The role of bacteriolysis in the pathophysiology of inflammation, infection and post-infectious sequelae, *APMIS.* 2002; 110:753–770.

[25] Glenny AT, Pope CG, Waddington H, and Wallace U. Immunological notes. XXIII. *The antigenic value of toxoid precipitated by potassium alum.* 1926;38 – 45.

[26] Golenbock DT, Hampton RY, Qureshi N, Takayama K, Raetz CR. Lipid-A-like molecules that antagonise the effects of endotoxins on human monocytes. *J. Biol. Chem.* 1991;266:19490–19498.

[27] Golkar M, Shokrgozar MA, Rafati S, Musset K, Assmar M, Sadaie R, Cesbron-Delauw, MF, and Mercier, C. *Evaluation of protective effect of recombinant dense granule antigens GRA2 and GRA6 formulated in monophosphoryl Lipid-A (MPL) adjuvant against Toxoplasma chronic infection in mice.* 2007;4301 – 4311.

[28] Guo L, et al. *Lipid-A acylation and bacterial resistance against vertebrate antimicrobial peptides.*1998; 95,189–198.

[29] Hall MA, et al. Intranasal immunisation with multivalent group A streptococcal vaccines protects mice against intranasal challenge infections. *Infect. Immun.* 2004;72:2507–2512.

[30] Hardaway RM, A review of septic shock, *Am. Surg.* 2000;66: 22–29.

[31] Hirschfeld M, Ma Y, Weis, JH, Vogel SN, and Weis J Cutting edge: repurification of lipopolysaccharide eliminates signaling through both human and murine toll-like receptor. *J. Immunol.*2000;165:61 – 622.

[32] Hopkin DA, Frapper fort ou frapper doucement: a gram-negative dilemma, *Lancet* 2. 1978; 1193–1194.

[33] Jacobson R M, Vaccine safety. *Immunol. Allergy Clin. North Am.* 2003: 23, 589 – 603.

[34] Jennings HJ, Bhattacharjee AK, Kenne L, et al. The R-type lipopolysaccharides of Neisseria meningitidis. *Can J Biochem* 1980;58(2):128-36.

[35] Jordan MB, Mills DM, Kappler J, Marrack P, Cambier JC. Promotion of B cell immune responses via an alum-induced myeloid cell population. *Science* 2004; 304:1808–1810.

[36] Kawai T, Akira S: TLR signaling. *Cell Death Differ.* 2006; 13: 816–825.

[37] Kemper C. and Atkinson JP. T-cell regulation: with complements from innate immunity. *Nat. Rev. Immunol.* 2007;7:9 – 18.

[38] Kidon MI, Shechter E, Toubi E. Vaccination against human papilloma virus and cervical cancer. *Harefuah.* 2011;150:33–36.

[39] Kong Q, Six DA, Roland KL, et al. Salmonella synthesising 1-dephosphorylated lipopolysaccharide exhibits low endotoxic activity while retaining its immunogenicity. *J Immunol* 2011;187(1):412-23.

[40] Kool M, Soullie T, van NM, Willart MA, Muskens F, Jung S, Hoogsteden HC, Hammad H, Lambrecht BN. Alum adjuvant boosts adaptive immunity by inducing uric acid and activating inflammatory dendritic cells 1. *J Exp Med.* 2008.

[41] Kotani S, et al. Synthetic Lipid-A with endotoxic and related biological activities comparable to those of a natural Lipid-A from an *Escherichia coli* re-mutant. *Infect. Immun.* 1985;49:225–237.

[42] Labadie J. Postlicensure safety evaluation of human papilloma virus vaccines. *Int. J. Risk Saf. Med.* 2011;23:103–112.

[43] Lawson LB, Norton EB, Clements JD. Defending the mucosa: adjuvant and carrier formulations for mucosal immunity. *Curr. Opin. Immunol.* 2011;23:414–420.

[44] Le Dur A, Caroff M, Chaby R, et al. A novel type of endotoxin structure present in Bordetella pertussis. Isolation of two different polysaccharides bound to Lipid-A. *Eur J Biochem.* 1978;84(2):579-89 25.

[45] Lee MS and Kim YJ. Signaling pathways downstream of pattern-recognition receptors and their cross talk. *Annu. Rev. Biochem.* 2007;76:447 – 480.

[46] Li C, Guan Z, Liu D, et al. Pathway for Lipid-A biosynthesis in Arabidopsis thaliana resembling that of *Escherichia coli*. Proc Natl Acad Sci USA. 2011;108(28):11387-92.

[47] Llanos-Cuentas A, et al. A clinical trial to evaluate the safety and immunogenicity of the LEISH-F1 + MPL-SE vaccine when used in combination with sodium stibogluconate for the treatment of mucosal leishmaniasis. *Vaccine.* 2010;28:7427–7435.

[48] Low KB, Ittensohn M, Le T, et al. Lipid-A mutant Salmonella with suppressed virulence and TNFalpha induction retain tumor-targeting in vivo. *Nat Biotechnol* 1999;17(1):37-41.

[49] Marohn ME, Barry EM. Live attenuated tularemia vaccines: recent developments and future goals. *Vaccine* 2013;31(35):3485-91.

[50] Mcaleer JP, Vella AT. Understanding how lipopolysaccharide impacts CD4 T-cell immunity. *Crit Rev Immunol* 2008;28(4): 281-99.

[51] McKee AS, Munks MW, Marrack P. How do adjuvants work? Important considerations for new generation adjuvants 1. *Immunity* 2007; 27:687–690.

[52] Miggin SM, O'Neill, LA. New insights into the regulation of TLR signaling. *J. Leukoc. Biol.* 80. 2006; 220–226.

[53] Miller SI, Ernst RK, Bader MW. LPS, TLR4 and infectious disease diversity. *Nat. Rev. Microbiol.* 2005;3:36–46.

[54] Moffatt JH, Harper M, Harrison P, et al. Colistin resistance in Acinetobacter baumannii is mediated by complete loss of lipopolysaccharide production. *Antimicrob Agents Chemother* 2010;54(12):4971-7.

[55] Morello CS, Levinson MS, Kraynyak KA, Spector DH. Immunisation with herpes simplex virus 2 (HSV-2) genes plus inactivated HSV-2 is highly protective against acute and recurrent HSV-2 disease. *J. Virol.* 2011;85:3461–3472.

[56] Morrison DC, Kline LF: Activation of the classical and properdin pathways of complement by bacterial lipopolysaccharides (LPS). *J. Immunol.* 1977; 118:362–368.

[57] Muta T, and Takeshige K. Essential roles of CD14 and lipopolysaccharide-binding protein for activation of toll-like receptor (TLR)2 as well as TLR4 Reconstitution of. *Eur. J. Biochem.* 2001;268: 4580 – 4589.

[58] O'Hagan, DT. MF59 is a safe and potent vaccine adjuvant that enhances protection against influenza virus infection. *Expert Rev. Vaccines.* 2007;6:699–710.

[59] Olson, K, et al. Liposomal gD ectodomain (gD1- 306) vaccine protects against HSV2 genital or rectal infection of female and male mice. *Vaccine.* 2009;28:548–560.

[60] Palsson-McDermott EM, O'Neill LA, Signal transduction by the lipopolysaccharide receptor, Toll-like receptor-4, *Immunology.* 2004; 113:153–162.

[61] Peng D, Hong W, Choudhury BP, et al. Moraxella catarrhalis bacterium without endotoxin, a potential vaccine candidate. *Infect Immun* 2005;73(11):7569-77.

[62] Piet JR, Zariri A, Fransen F, et al. Meningitis caused by a lipopolysaccharide deficient Neisseria meningitidis. *J Infect.* 2014;69(4):352-7.

[63] Polhemus ME, et al. Evaluation of RTS, S/AS02A and RTS, S/AS01B in adults in a high malaria transmission area. *PLoS One.* 2004;4:e6465.

[64] Poltorak A, He X, Smirnova I et al., Defective LPS signaling in C3H/HeJ and C57BL/10ScCr mice: mutations in Tlr4 gene, *Science.* 1998; 282:2085–2088.

[65] Poltorak A, He X, Smirnova I, et al. Defective LPS signaling in C3H/HeJ and C57BL/10ScCr mice: mutations in Tlr4 gene. *Science* 1998;282(5396):2085-8.

[66] Pulendran B. and Ahmed R. Translating innate immunity into immunological memory: implications for vaccine development. *Cell.* 2006;124:849 – 863.

[67] Qureshi ST, Lariviere L, Leveque G, Clermont S, Moore KJ, Gros P, and Malo D. Endotoxin-tolerant mice have mutations in Toll-like receptor 4 (Tlr4*). J. Exp. Med.* 1999; 189:615 – 625.

[68] Raetz CR, Reynolds CM, Trent MS, et al. Lipid-A modification systems in gram-negative bacteria. *Annu Rev Biochem* 2007;76:295-329.

[69] Raetz CR, Whitfield C. Lipopolysaccharide endotoxins. *Annu Rev Biochem* 2002;71: 635-700.

[70] Rietschel ET, Brade H, Holst O, Brade L, Muller-Loennies S, Mamat U, Zahringer U, Beckmann F, Seydel U, Brandenburg K, Ulmer AJ, et al., Bacterial endotoxin: chemical constitution, biological recognition, host response, and immunological detoxification, *Curr. Top. Microbiol. Immunol.* 1996; 216: 39–81.

[71] Rietschel ET, Kirikae T, Schade FU, et al. The chemical structure of bacterial endotoxin in relation to bioactivity. *Immunobiology.* 1993;187(3-5):169-90.

[72] Robinson MJ, Sancho D, Slack EC, LeibundGutLandmann S, and Reis e Sousa. Myeloid C-type lectins in innate immunity. *Nat. Immunol.* 2006;7:1258 – 1265.

[73] Saitoh S, et al. Lipid-A antagonist, lipid IVa, is distinct from Lipid-A in interaction with Toll-like receptor 4 (TLR4)-MD-2 and ligand-induced TLR4 oligomerisation. *Int. Immunol.* 2004;16:961–969.

[74] Schwandner R, Dziarski R, Wesche H, Rothe M, Kirschning CJ, Peptidoglycan- and lipoteichoic acid-induced cell activation is mediated by toll-like receptor 2, *J. Biol. Chem.* 1999; 274:17406–17409.

[75] Takada H, Kotani S. Structural requirements of Lipid-A for endotoxicity and other biological activities. *Crit Rev Microbiol.* 1989;16(6):477-523.

[76] Takeuchi O, Hoshino K, Kawai T, Sanjo H, Takada H, Ogawa T, Takeda K, and Akira S. Differential roles of TLR2 and TLR4 in recognition of gram-negative and grampositive bacterial cell wall components. *Immunity.* 1999;11:443 – 451.

[77] Tana WS, Isogai E, Oguma K. Induction of intestinal IgA and IgG antibodies preventing adhesion of verotoxin-producing *Escherichia coli* to Caco-2 cells by oral immunisation with liposomes. *Lett. Appl. Microbiol.* 2003;36,135–139.

[78] Trent MS, Pabich W, Raetz CR, Miller SI. A PhoP/PhoQ-induced Lipase (PagL) that catalyses 3-O-deacylation of Lipid-A precursors in membranes of Salmonella typhimurium. *J. Biol. Chem.* 2001. 276;9083–9092.

[79] Triantafilou M, Brandenburg K, Kusumoto S, Fukase K, Mackie A, Seydel U, and Triantafilou K. Combinational clustering of receptors following stimulation by bacterial products determines lipopolysaccharide responses. *Biochem. J.* 2004;381:527 – 536.

[80] Trinchieri G, Sher A. Cooperation of Toll-like receptor signals in innate immune defence. *Nat. Rev. Immunol.* 2007;7:179–190.

[81] Ulevitch RJ, Tobias PS. Recognition of gram-negative bacteria and endotoxin by the innate immune system, *Curr. Opin. Immunol.* 1999; 11, 19–22.

[82] Ulrich JT, Myers KR. Monophosphoryl Lipid-A as an adjuvant. Past experiences and new directions. *Pharm Biotechnol* 1995;6: 495-524.

[83] Van der Ley P, Steeghs L, Hamstra HJ, et al. Modification of Lipid-A biosynthesis in Neisseria meningitidis lpxL mutants: influence on lipopolysaccha-ride structure, toxicity, and adjuvant activity. *Infect Immun* 2001;69 (10):5981-90.

[84] Vernacchio, L., et al. Effect of monophosphoryl Lipid-A (MPL) on T-helper cells when administered as an adjuvant with pneumocococcal-CRM197 conjugate vaccine in healthy toddlers. *Vaccine.* 2002;20:3658–3667.

[85] Von EK. et al. The candidate tuberculosis vaccine Mtb72F/AS02A: Tolerability and immunogenicity in humans. *Hum. Vaccin.* 2009;5: 475–482.

[86] Wang X, Ribeiro AA, Guan Z, et al. *Attenuated virulence of a Francisella mutant lacking the Lipid-A 4'-phosphatase.* Proc Natl Acad Sci USA. 2007;104(10): 4136-41.

[87] Woltmann A, Hamann L, Ulmer AJ, et al., Molecular mechanisms of sepsis, *Langenbeck's Arch. Surg.* 1998; 383: 2–10.

[88] Zamze SE, Moxon ER. Composition of the lipopolysaccharide from different capsular serotype strains of Haemophilus influenzae. *J Gen Microbiol.* 1987; 133(6):1443-51.

[89] Zughaier SM, Zimmer SM, Datta A, et al. Differential induction of the toll-like receptor 4-MyD88-dependent and - independent signaling pathways by endotoxins. *Infect Immun.* 2005;73(5): 2940-50.

Chapter 6

Endotoxins in the Environment

Gurvinder Kaur[1,*], Arif Pandit[2], Saifun Nisa[3] and R S Sethi[4]

[1]College of Animal Biotechnology, Guru Angad Dev Veterinary and Animal Sciences University, Punjab, India

[2]Directorate of Research, Sher e Kashmir University of Agricultural Sciences and Technology of Kashmir, Kashmir. India

[3]Department of Environmental Science, School of Earth & Environmental Sciences, University of Kashmir. Srinagar, India

[4]Department of Animal Biotechnology, College of Animal Biotechnology, Guru Angad Dev Veterinary and Animal Sciences University, Punjab India

Abstract

Gram-negative bacteria such as Escherichia *coli, Pseudomonas luteola, Bacillus cereus* etc., are characterised by lipopolysaccharide (LPS) containing outer membrane and LPS varies in chemical structure in different Gram-negative species and even among strains of the same species. LPS has endotoxin activity, whose potency varies from species to species and is considered a potent pro-inflammatory molecule. Endotoxins are present ubiquitously in the indoor, occupational and outdoor environment due to constant shedding of the bacterial cell wall or bacterial lysis. There is a potential risk for animals, consumers, and workers to be exposed to possibly harmful amounts of Endotoxin during the production of various feed additives, such as amino acids and vitamins, by employing Gram-negative bacteria, particularly Escherichia coli in the livestock industry. Biomedical, livestock, agricultural, and food industries are at higher risk of endotoxin exposure and other

* Corresponding Author's E-mail: kaur.gurvinder2016@gmail.com.

In: Endotoxins and their Importance
Editors: Arif Pandit and R. S. Sethi
ISBN: 978-1-68507-839-3

xenobiotics. In this chapter, we will discuss the potential areas of endotoxin exposure in the environment and what measures can be adopted to lessen its ecological threat.

Keywords: LPS, Gram-negative bacteria, environment, risk factors

Introduction

One of the bacterial products that exists in close association with the bacterium and released upon bacteriolysis can elicit profound effects on humans and other animals. Bacterial endotoxins are the most frequently reported pyrogens. Their purified derivatives, lipopolysaccharides (LPS), act as potent inflammatory molecules (Michel 2000) and are highly toxic substances to other organisms. LPS is known as one of the most effective immune activators, causing non-specific immune reactions in the host tissues (Rauber et al. 2014). There is always a danger of substantial Endotoxin or LPS contamination in many different environments due to their ubiquitous presence and the shedding of the bacterial cell wall or bacterial lysis (Biomin 2016). LPS are chemically composed of an outer O-polysaccharide chain (40 repeat units), an inner R-polysaccharide chain and lipid A (Disaccharide phosphate and Fatty acid chains). The latter is the toxic part of the molecule. Levels of toxicity depend on this lipid A portion of Endotoxin, which acts as a trigger to elicit the immune response and varies in different species of bacteria. In addition, the Lipopolysaccharide layer on the outermost surface of the bacterium gives the maximal opportunity to interact with the external environment, and the most suitable example is the intestinal tract of most mammalian species.

Besides Endotoxins, numerous natural and manufactured toxins are present in the environment. Lead, Mercury, Radon, Formaldehyde, Benzene and cadmium are natural toxins and BPA (Bisphenol A), phthalates and pesticides are examples of manufactured toxins. All these toxins in high doses can negatively affect human health and can lead to fatality. Toxicity due to some may lead to deadly diseases like cancer, some can act as endocrine disruptors, and some may cause organ failure or developmental problems. Toxins from different microorganisms are also present in the environment like bacterial toxins and mycotoxins (produced from fungi). In this we will be more focused on the bacterial toxins, especially endotoxins which are a constant threat to number of niches.

6.1. Bacterial Toxins

Bacterial toxins are of two types of exotoxins and endotoxins. Exotoxins are secreted mainly by Gram-positive bacteria and are mainly proteinaceous. As the name suggests, Exotoxins are produced inside the pathogenic bacteria and secreted or released in the surrounding medium upon cell lysis and endotoxins as the component of the gram-negative bacteria are liberated when the bacteria die, and cell wall breaks apart. Toxins produced by Staphylococcus aureus, Bacillus cereus, Streptococcus pyogenes, Bacillus anthrcis (Alpha-toxin, also known as alpha-hemolysin (Hla)) are exotoxins and toxins produced by E. coli, Salmonella Typhi, Shigella, Vibrio cholera (Cholera toxin- also known as choleragen) are few examples of endotoxins.

Endotoxin is found in Gram-negative bacteria and bacterial products or debris. Many different environments may be contaminated with bacterial endotoxins or lipopolysaccharides (LPS) characterised by their high ubiquity level. Thus, Endotoxin is widely present in the environment, including dust, animal waste, foods, and other materials generated from, or exposed to, Gram-negative bacterial products. Accordingly, individuals involved in occupations associated with intensive livestock and/or agricultural operations are frequently exposed to elevated levels of Endotoxin. Given its prevalence in the environment, Endotoxin is also generally present endogenously in sites such as the airway and the gastrointestinal tract. In fact, endotoxin presentation as a bioaerosol is an important route of occupational exposure associated with the development and progression of airway disease. In addition, mechanical- or pathogen-induced (e.g., virus) injury may allow for secondary bacterial introduction and thus lead to endotoxin presentation to the tissues or blood, whereupon it initiates a robust immunological response.

6.2. Endotoxins in Different Environments

6.2.1. Endotoxins in Biomedical Industry

Biomedical research has been expanding over the last decade to provide improved medical devices in science, and endotoxins are named an invisible companion in biomaterials research. In the quest to provide the best product, several manufacturers and non-chemists do not understand the complex interaction mechanisms of such natural toxins. Medical implants and polymeric devices for the application in the clinical treatment of orthopaedic tissue injuries are increasingly coated with bioactive biomaterials derived from natural substances to induce desirable biological effects. Many metals and polymers used in biomaterials research show a high affinity for endotoxins, abundant in the environment. Endotoxin contamination is indicated in the pathology of periodontitis and aseptic implant loosening, which may affect the orthopaedic fixation and may also affect the evaluation of a biomaterial's bioactivity by inducing strong inflammatory reactions. Chitosan, the partly deacetylated form of chitin, is a metal implant coated with bioactive materials to improve osseointegration and performance at the bone–biomaterial interface. Lieder et al. (2013) have shown the presence of bacterial endotoxins in chitosan derivatives, resulting in false-positive results, profoundly altering product performance in in-vitro assays. LPS in high concentration can cause septic shock and acute renal failure in humans (Yoon et al. 2007). Therefore, strict regulations need to be adopted to decrease the risk of endotoxin contamination in medical preparations and avoid further discrepancies.

Biofilms are organised layers of microbes that can coat any surface. We can find a biofilm in everyday objects (bags, toothbrushes, etc.), for instance, in food or medical supplies. Gram-negative bacteria develop on the biofilm and can be a significant source of contamination with endotoxins. Haemodialysis is one such procedure performed in patients with kidney failure, which often causes infections in these people, mainly due to the bacterial biofilm that can be found in used medical equipment. Gram negative bacteria are responsible for most cases of the diseases contracted during haemodialysis. Within a biofilm, the behaviours of organisms are often different compared with the non-biofilm state (particularly planktonic cells) as a result of different genes being turned on or turned off. Modification to the Lipopolysaccharide molecule within a biofilm can occur by incorporating a palmitate acyl chain into the lipid A part of lipopolysaccharide and can lead to different reactions like enhanced inflammatory cytokine response, as shown with Pseudomonas aeruginosa strains (Ciornei et al. 2010). There is also evidence that surface-associated endotoxin levels correlate with biofilm levels, which also serves as a detection method for the presence of biofilms (Donlan and Costerton, 2002).

In *the pharmaceutical industry, endotoxins pose a risk to many drugs manufacturing processes* and finished products. In the pharmaceutical industry, endotoxin testing of parenteral drugs is vital because lipopolysaccharide is ubiquitous in waterborne bacterial species, and water is the main ingredient in many parenteral products and is also used for cleaning all kinds of glassware used in research e.g., a single *Escherichia coli* cell contains approximately 2 million lipopolysaccharide molecules (Caroff et al. 2002). These molecules consist of a hydrophobic lipid A moiety, a complex array of sugar residues and negatively charged phosphate groups. Endotoxins are heat stable, making them resistant to most conventional sterilisation processes and thus necessitating separate tests for viable cells (bioburden) and Endotoxin. Endotoxin exposure can also stem from human handling, dust, packaging, contaminated rinse water, microbial growth (Williams 2001) and processes involved, for example, Gram-negative bacteria, such as *E. coli*, are often used to produce recombinant DNA products like peptides and proteins. These products are always contaminated with endotoxins, and removal steps are required to eliminate the Endotoxin from the product (Hirayama and Sakata 2002). Two difficulties relate to endotoxin removal from products. The first is that the process applied must not alter the product during endotoxin clearance. The second is with the relatively low endotoxin concentration in the presence of the product and the difficulty in removing bound Endotoxin. The binding of Endotoxin can become enhanced through protein concentration processing steps.

The emergence of multiple drug-resistant (MDR) bacteria has led to patients receiving multiple drugs at higher doses with a constant threat to potential endotoxin exposure on the patient. As medical technology improves, many at-risk patients, such as the immunosuppressed, premature babies and the elderly, with increased sensitivity to pyrogens.

Regularly used laboratory equipments are at higher risk of being contaminated with endotoxins. The staff working in research laboratories or production must have specific control measures when working with products that must not be contaminated with any type of toxins for their subsequent application. For example, among the materials that are susceptible to be contaminated with LPS are nanoparticles, which have been used in medicine for some time. It has been found that most nanomaterials contain endotoxins, which may even hinder the effectiveness of the treatment in Nanomedicine. The control regarding the levels of endotoxins when it comes to parenteral vaccines is a clear example of the importance of contamination with these toxins on the health of animals and humans.

6.2.2. Endotoxins in Livestock Industry

Livestock involves raising the animals like cattle, swine, sheep, horses, and to a lesser extent, goats and mules, and the processing of the animal products for consumers. Livestock provides food items such as Milk, Meat and Eggs and contributes to the production of wool, hair, hides, and pelts. Dung and other animal wastes serve as excellent farmyard manure and the value of it is worth several crores of rupees. Diseased or Susceptible animals are more prone to endotoxin exposure. There are many reasons of endotoxin exposure to livestock, like the transport of livestock, which sometimes turn out to be the most stressful and injurious stage in the chain of operations between farm and target spots resulting in poor animal welfare and loss of production. Modern and commonly used dairy breeds like the Holstein Friesian are developed in northern countries, so they are tolerant towards cold weather conditions but more susceptible to long hot summers, which leads to heat stress. Each year over US$ 1 billion is lost due to heat stress in cattle in the United States alone. Annual losses in dairy and beef are estimated to be $ 897 million and $ 369 million, respectively (Dairy Global, 2019).

Heat stress can disrupt the rumen, harm the gut integrity and affects several physiological functions. The disruption of the rumen or sub-acute rumen acidosis (SARA) and impaired gut barrier can lead to the translocation of toxins (e.g., endotoxins, mycotoxins) as well as unwanted metabolites (e.g., biogenic amines such as histamine) into the blood circulation and generates a robust inflammatory response. Endotoxins bind to TLR-4 receptors and upregulate the cytokines like IL-6, TNF- α and IL-1β. This Inflammatory cascade triggers the acute phase proteins e.g., haptoglobin LPS-binding protein (Nicole 2020). This acute phase response consumes a lot of energy, which the animal cannot use anymore for growth or milk production. In the worst case, high amounts of endotoxins can result in septic shock and even death, also defined as Endotoxemia (All about feed, 2016). Furthermore, when LPS compromise intestinal integrity, the risk of secondary infections increases and production performance may decline. A recent study from the research group at Iowa State University showed that the *in vivo* administration of endotoxins drastically decreased the milk yield in Holstein cows by 80%.

When endotoxins are released into the intestinal lumen of chickens or swine or in the rumen of polygastric animals, they can cause severe *damage to the animal's health and performance* by over-stimulating its immune system. The toxicity of LPS is mainly caused by lipid A; however, both lipid A and O-antigen of LPS stimulate the immune system. This happens when it passes the mucosa and enters the bloodstream or attacks the leukocytes. The intestinal mucosa is almost impermeable to external agents but there are some vulnerable points (Zachary 2017) where endotoxins can be infused like cells of the *lamina propria* (a layer of connective tissue underneath the epithelium) through the microfold (M) cells of the Peyer's patches (which consist of gut-associated lymphoid tissue). The M cells are not covered by mucus and thus exposed. LPS can also encounter the lymphocytes or reach the lymph nodes through the afferent lymphatic vessels, lastly, they might affect the tight junctions, the multiprotein complexes that keep the enterocytes (cells that form the intestinal villi) cohesive. LPS can break the tight junctions, reaching the first capillaries and, consequently, the bloodstream by destabilising the protein structures and triggering enzymatic reactions that chemically degrade them. LPS can also pass through the mucosa, become entangled and meet the lymphocytes or can reach the regional lymph nodes. Lastly, LPS might affect the tight junctions, the multiprotein complexes that keep the enterocytes (cells that form the intestinal villi) cohesive. LPS can break the tight junctions, reaching the first capillaries and, consequently, the bloodstream by destabilising the protein structures and triggering enzymatic reactions that chemically degrade them.

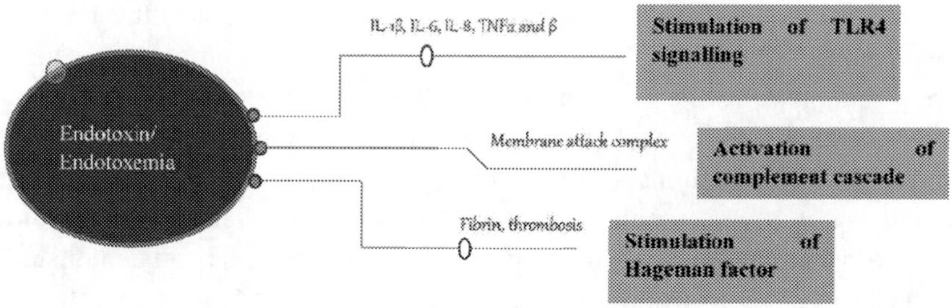

Figure 6.1. How Endotoxin causes endotoxemia- modes of action.

The inflammation response can result in mitochondrial injury to the intestinal cells, which alter the cellular energy metabolism. This is reflected in changes to the adenosine triphosphate (ATP) levels, the energy "currency" of living cells. A study by Li et al., (2015) observed *a respective reduction of 15% and 55% in the ATP levels of the jejunum and ileum of LPS-challenged broilers,* compared to the unchallenged control group. This depicts the extent to which animals lose energy while they experience (more or less severe) endotoxemia. One more study by Huntey et al., (2017) found that LPS-challenged pigs retained 15% less of the available metabolisable energy and showed 25% less nutrient deposition. These results show concrete metabolic consequences caused by the febrile response to endotoxemia. Endotoxin mitigation is essential to mitigate the losses that occurred due to endotoxemia and septic shock. EW Nutrition's Mastersorb Gold is an anti-mycotoxin agent, and its components effectively bind bacterial toxins.

Another main reason of endotoxin exposure is through the feed additives. Livestock are frequently exposed to a relatively high content of Endotoxin in the diet. The workers processing a dusty additive may also be exposed to hazardous amounts of Endotoxin even if the endotoxin concentration of the product is low. Farmworkers and workers in the premixture factory, i.e., where minerals/vitamins/trace nutrients supplements are prepared, are more susceptible to endotoxins arising from feed additives and animal faeces. (Health council 2010).

6.2.3. Endotoxins in the Poultry Sector

Endotoxin exposure can impair performance and impact individual poultry birds differently depending upon various factors. Poultry are continuously exposed to lipopolysaccharides throughout their lives. In healthy birds, the intestinal and other epitheliums such as skin or lungs represent an effective barrier that prevents the passage of lipopolysaccharides into the bloodstream. Once there, however, endotoxins can elicit strong immune responses, weakening birds' immune systems and impairing performance. The severely pronounced immune response can lead to septic shock. This activation of the immune system, along with increased oxidative stress, has been demonstrated in chicken cells in vitro. After incubation with different concentrations of lipopolysaccharides (20 and 40 ng/mL), a chicken macrophage cell line (HD11) showed increased oxidative stress, as demonstrated by enhanced levels of nitrite oxide (NO) production (Li et al. 2015). In recent years, new concerns have emerged about nutritional, environmental and social factors that may disrupt the barrier function and increase exposure to lipopolysaccharides. Such exposure may result in clinical or sub-clinical signs that ultimately affect poultry production.

Many other factors like dietary changes, high temperature, presence of mycotoxins in feed, and use of antibiotics contribute to endotoxin exposure. Dietary changes play a clear role. For example, moving birds from a corn-based diet to a rye-wheat-barley diet results in increased lipopolysaccharide levels in blood serum along with an increase in inflammatory markers. Mycotoxins such as deoxynivalenol are very well known to disrupt the intestinal barrier. According to the latest Biomin Mycotoxin Survey results, the presence of deoxynivalenol, zearalenone and fumonisins in finished feed have all increased in recent years. Additionally, sub-therapeutic use of antibiotics in some countries raises concerns about antibiotic resistance and the change in gut microflora and the plausible release of lipopolysaccharides in the gut lumen.

6.2.4. Endotoxins in Agricultural Industry

Agriculture is the primary source of livelihood for about 60% of India's population, and demand for agricultural inputs and allied services like warehousing and cold storage is increasing in India at a fast pace (IBEF 2020). Modern agronomy, which enables agrochemicals like pesticides and fertilisers and various technological advancements, has sharply increased yields but poses widespread ecological and environmental threats. (Skórska 2005). Workers involved in cultivating, harvesting, storing or processing of agricultural products may be exposed, via the respiratory route, to a wide range of airborne biological agents such as microorganisms, Endotoxin, peptidoglycan, glucans, lipoteichoic acid, and allergens. Farmers working with cotton are highly exposed to cotton dust, often contaminated with gram-negative bacteria containing endotoxins (Paudyal et al. 2011). Mukherjee et al. (2004) in a study conducted on Indian jute mill workers, found that increased exposure to bacterial Endotoxin in airborne dust causes respiratory problems.

Some researchers have reported that LPS, in synergy with other substances present in organic dust or with pesticides, plays an essential role in airway inflammation and bronchoconstriction. High occupational endotoxin exposure via inhalation is prevalent in agricultural and related industries. The study reported that lindane and indoxacarb, along with co-exposure with endotoxins, causes changes in lung morphology and alters TLR-4 and TNF-α (Pandit et al. 2016) and TLR-9 expression, respectively (Kaur et al. 2016). Various studies shows that LPS interaction with various classes of pesticides modulates pulmonary responses during pesticide-induced lung insult (Pandit et al. 2017 and Pandit et al. 2019; Sethi et al. 2017; Verma et al. 2018, Kaur et al. 2021a, Kaur et al. 2021b).

6.2.5. Endotoxins in Food Industry

The incidence of food-borne infections has increased globally with many human populations at risk (Cho et al. 2011) and Gram-negative and Gram-positive bacteria pathogens are at the epicentre of most reported cases (Sudershan et al. 2014). Apart from the fact that Gram-negative bacteria in fermented foods may cause infections, they can be toxigenic, producing endotoxins in foods. The potency of Endotoxin varies amongst different bacterial species, but E. coli produces LPS with extremely high endotoxin activity hence it is often used as the model organism (Raetz and Whitfield, 2002). The endotoxins present in various foods can be associated with several factors affecting the quality of these food products along the food chain. Exposure to these harmful toxins can occur at any stage and Poor food safety knowledge, use of contaminated raw materials, utilisation of polluted water, inadequate hygienic practices, unstandardised production processes, mixed-culture processing, deplorable hygiene status of processing environments, poor packaging, inadequate preservation techniques, are some of the factors that provoke the presence of these pathogenic organisms in foods. The Indian food processing industry accounts for 32% of the country's total food market, one of the largest industries in India and is ranked fifth in terms of production, consumption, export and expected growth. Therefore, to reduce the presence of Gram-negative bacteria and their toxins in food products, proper assessment tests should be carried out, and further development and enactment of adaptable food safety measures should be done.

With regard to the pathogenesis of baker's asthma, for many years it was believed that IgE mediated the response. This was considered the main and exclusive mechanism by which asthma developed. However, it has been suggested that the innate immune response may contribute to the development of baker's asthma as wheat flour contains bacterial endotoxins or lipopolysaccharides (LPS) that play an important role in the development of asthma. We can list wheat flour among the environments where we can find bacterial endotoxins. Other workers are also exposed to contamination with endotoxins, such as those who work with livestock. The reason is that Gram-negative bacteria are increasingly producing the additives used in livestock feed. Therefore, there is the possibility of potential endotoxin exposure to the workers handling the additive, the consumer, and the target animals. The health council of the Netherlands proposed a recommended exposure limit of 90 IU/m3 for endotoxins in the workplace. Therefore, the exposure limit should be maintained lower than this limit and a threshold of 10 ng/m3 for an 8hr working day should be applied to prevent lung inflammation.

Conclusion

Endotoxins are widely present in the environment, including dust, animal waste, foods, and other materials generated from, or exposed to, Gram-negative bacterial products. Accordingly, individuals involved in occupations associated with intensive livestock and agricultural operations are frequently exposed to elevated levels of Endotoxin. Exposure to endotoxins may be associated with a compromised immune state and other health problems therefore, strict regulations and practices should be adopted to minimise its exposure in different environments.

References

Biomin. 2016. *Endotoxins and their negative impact on poultry*. www.biomin.net.

Caroff M, Kariban D, Cavaillon J. et al., 2004. Structural and functional analyses of bacterial lipopolysaccharides. *Microbes and Infection*. 4: 915-926.

Cho J I, Lee S H, Lim JS. et al., 2011. Detection and distribution of food-borne bacteria in ready-to-eat foods in Korea. Food Sci Biotechnol 20: 525.

Ciornei C D, Novikov A, Beloin C, Fitting C, Caroff M, Ghigo J M, Cavaillon J M, Adib-Conquy. M. 2010. Biofilm-forming Pseudomonas aeruginosa bacteria undergo lipopolysaccharide structural modifications and induce enhanced inflammatory cytokine response in human monocytes. *Innate Immun.* 16:288–301.

Donlan R M and Costerton J W. 2002. Biofilms: Survival Mechanisms of Clinically Relevant Microorganisms, *Clin. Microbiol. Rev.* 15 (2): 167-193.

Health-based recommended occupational exposure limit. The Hague: Health Council of the Netherlands; 2010. Publication no. 2010/04OSH Health Council of the Netherlands. *Endotoxins* 2010.

Hirayama C, Sakata M. 2002. Chromatographic removal of Endotoxin from protein solutions by polymer particles. *J. Chrom. B.* 781: 419-432.

Huntley, Nichole F, C. Martin Nyachoti, and John F. Patience. 2017. *Immune System Stimulation Increases Nursery Pig Maintenance Energy Requirements. Iowa State University Animal Industry Report* 14(1).

Kaur G, Verma R, Mukhopadhyay C S and Sethi R. 2021. Elevated pulmonary levels of Axin2 in mice exposed to herbicide 2, 4-D with or without Endotoxin. *Journal of Biochemical and Molecular Toxicology*, e22912.

Kaur Geetika, Sunil B V, Singh Baljit, Sethi R S. 2021. Exposures to 2,4-Dichlorophenoxyacetic acid with or without Endotoxin upregulate small cell lung cancer pathway. *J. Occup. Med. Toxicol.* 16:14.

Kaur S, Mukhopadhyay C S, Sethi R S. 2016. Chronic exposure to indoxacarb and pulmonary expression of toll-like receptor-9 in mice. *Vet World.* 9: 1282.

Li Jiaolong, Yongqing Hou, Dan Yi, Jun Zhang, Lei Wang, Hongyi Qiu, Binying Ding, and Joshua Gong. 2015. Effects of Tributyrin on Intestinal Energy Status, Antioxidative Capacity and Immune Response to Lipopolysaccharide Challenge in Broilers. *Asian-Australasian Journal of Animal Sciences.* 28(12): 1784–93.

Lieder Ramona, Vivek S. Gaware, Finnbogi Thormodsson, Jon M. Einarsson, Chuen-How Ng, Johannes Gislason. et al., 2013. Endotoxins affect bioactivity of chitosan derivatives in cultures of bone marrow-derived human mesenchymal stem cells, *Acta Biomaterialia* 9(1): 4771-4778.

Michel O. 2000. Systemic and local airways inflammatory response to Endotoxin. *Toxicology*. 152(1-3): 25-30.

Mukherjee, Ashit & Chattopadhyay, Bhakar & Bhattacharya, Sanat & Saiyed, Habibullah. 2004. Airborne Endotoxin and Its Relationship to Pulmonary Function among Workers in an Indian Jute Mill. *Archives of environmental health.* 59: 202-8.

Nicole Reisinger, Caroline Emsenhuber, Barbara Doupovec, Elisabeth Mayer, Gerd Schatzmayr, Veronika Nagl, and Bertrand Grenier. 2020. Endotoxin Translocation and Gut Inflammation Are Increased in Broiler Chickens Receiving an Oral Lipopolysaccharide (LPS) *Bolus during Heat Stress Toxins.* 12(622).

Pandit AA, Choudhary S. Ramneek, Singh B, and Sethi RS. 2016. Imidacloprid induced histomorphological changes and expression of TLR-4 and TNF-α in the lung. Pesticide Biochemistry and Physiology 131: 9–17

Pandit A A, Mukhopadhyay C S, Sethi R S. 2017. Expression of TLR-9 and IL-1 following concomitant exposure to Imidacloprid and Endotoxin. *Pestic. Res. J.* 29: 243–50.

Pandit AA, Gandham RK, CS Mukhopadhyay, Ramneek, and RS Sethi. 2019. Transcriptome analysis reveals the role of the PCP pathway in fipronil and endotoxin-induced lung damage. Respiratory Research. 20:24

Paudyal Priyamvada, Semple Sean, Niven Robert, Gael Tavernier, Jonathan G. Ayres. 2011. Exposure to Dust and Endotoxin in Textile Processing Workers. *The Annals of Occupational Hygiene.* 55(4): 403–409.

Raetz C R H, Whitfield C 2002. Lipopolysaccharide endotoxins. *Annu. Rev. Biochem.* 71, 635-700.

Rauber R, Perlin V, Fin C, Mallmann A, Miranda D, Giacomini L. & do Nascimento V. 2014. Interference of Salmonella typhimurium lipopolysaccharide on performance and biological parameters of broiler chickens. *Revista Brasileira de Ciencia Avícola.* 16: 77–81.

Sethi R S, Schneberger D, Charavaryamath C Singh B. 2017. Pulmonary innate inflammatory responses to agricultural occupational contaminants. *Cell Tissue Res.* 367: 627–642.

Skorska C, Sitkowska J, Krysinska-Traczyk E, Cholewa G, Dutkiewicz J. 2005. Exposure to airborne microorganisms, dust and Endotoxin during processing of peppermint and chamomile herbs on farms. *Ann. Agric. Environ. Med.* 12: 281-288.

Sudershan R V, Naveen R, Kashinath L, Bhaskar V, and Polasa K. 2014. *Foodborne Infections and Intoxications in Hyderabad India* Volume 2014 |Article ID 942961.

Verma G, Mukhopadhyay C S, Verma R Singh B. Sethi R S. Long-term exposures to ethion and endotoxin cause lung inflammation and induce genotoxicity in mice. *Cell Tissue Res.* 2018.

Wallace R John, Gropp Jurgen, Dierick Noel, Costa G Lucio, Martelli Giovanna et al., 2016. *Risks associated with endotoxins in feed additives Environmental health.* 15:5.

Williams K L. 2002. Endotoxins: Pyrogens, LAL testing and Depyrogenation 2nd Edition. *Drugs and the Pharmaceutical Sciences* Volume 111. Marcel Dekker Inc., New York, USA. Chapters 1, 2, 7 & 8.

Yoon H J, Moon M E, Park H S, Im S Y, Kim Y H. 2007. Chitosan oligosaccharide (cos) inhibits LPS-induced inflammatory effects in RAW 264.7 macrophage cells. *Biochem. Biophys. Res. Commun.* 358:954–9.

Zachary, James F. 2017. "Chapter 4 – Mechanisms of Microbial Infections." Essay. In *Pathologic Basis of Veterinary Disease*, 132–241. St Louis, MO: Mosby, 2017.

About the Editors

Dr. Arif Pandit, PhD

Subject Matter Specialist and Junior Scientist (Animal Science)
Sher e Kashmir University of Agricultural Sciences
and Technology of Kashmir
Shalimar, Srinagar
Jammu and Kashmir, India

Dr. R. S. Sethi, PhD

Professor and Head
Department of Animal Biotechnology
College of Animal Biotechnology
Guru Angad Dev Vety and Animal Sciences University
Punjab, India

Index

M

N

O

P

R

S

T

V

W